Boos
Sho

Booster Shots

The Urgent Lessons of Measles and the Uncertain Future of Children's Health

Adam Ratner, MD, MPH

AVERY
an imprint of Penguin Random House
New York

AVERY

an imprint of Penguin Random House LLC
penguinrandomhouse.com

Copyright © 2025 by Adam Ratner

Penguin Random House values and supports copyright. Copyright fuels creativity, encourages diverse voices, promotes free speech, and creates a vibrant culture. Thank you for buying an authorized edition of this book and for complying with copyright laws by not reproducing, scanning, or distributing any part of it in any form without permission. You are supporting writers and allowing Penguin Random House to continue to publish books for every reader. Please note that no part of this book may be used or reproduced in any manner for the purpose of training artificial intelligence technologies or systems.

Most Avery books are available at special quantity discounts for bulk purchase for sales promotions, premiums, fund-raising, and educational needs. Special books or book excerpts also can be created to fit specific needs. For details, write SpecialMarkets@penguinrandomhouse.com.

Hardcover ISBN: 9780593330869
Ebook ISBN: 9780593330876

Printed in the United States of America
1st Printing

Book design by Ashley Tucker

For Shari and Samantha, always

It all depends on whose baby has the measles.
—FRANKLIN DELANO ROOSEVELT

Contents

Introduction xi

1.
Meet the Measles 1

2.
Storytelling 9

3.
Panum's Map 17

4.
Contagion in the Service of Empire 37

5.
Crowded, Poor, Malnourished 53

6.
Making Nothing Happen 71

7.
Vaccines Don't Save Lives. Vaccinations Save Lives. 99

8.
Imperfect Tools 113

9.
Amnesia 143

10.
Banner Years 165

11.
Booster Shots 183

Acknowledgments 203 • Notes 207 • Index 255

Introduction

I am a pediatrician.

When I was about fifteen years old, my brother Gabriel, a toddler, fell into the swimming pool in our backyard. I was in my bedroom—bored, annoyed—working through a review book for an upcoming standardized test in biology. My mother was unloading groceries. She had unbuckled Gabe from his car seat and thought that he was following her into the house. Instead, he wandered around the side, up a short set of stairs, and into the water. My room had a window that faced the back of the house, but I didn't hear him fall in—no one did. I hadn't even realized that they were home. This was a moment when some safeguards failed—a couple of minutes in the course of a lifetime. An unlatched gate, a distracted adult, an older brother who wasn't listening. My sister, only a few years older than Gabe, found him floating on the water—pale skin, bloated diaper.

For a lot of families, the story ends there. Attention lapses, a safety plan fails, and a child is lost. I am grateful to say that we were luckier than that. My mother, trained as both a pediatrician and an anesthesiologist, pulled him from the pool. Drawn by my sister's cries, I arrived, wanting to help but shaking and confused. I had never shook before. Mom gave orders—I called the emergency number, explained the unimaginable, blurted our address. I have no memory of what I said, but I do remember doing chest compressions—one hand on a minuscule chest—as she breathed for him. It was my first time

INTRODUCTION

seeing—much less helping to do—CPR outside the border of a television screen.

This is where a story of bad luck becomes a story of good luck. An ambulance arrived. Gabe went first to a local hospital and then, on a ventilator, to a children's hospital in the city. He spent weeks in their pediatric intensive care unit, a mammoth crowded room packed with a strange juxtaposition of beeping technology and mismatched lengths of tubing strung together with pink latex tape. A team kept careful track of Gabe, and he held up his end of the bargain—slowly healing, needing less support from machines as he went along. I visited him there, held his hand, and watched his protectors.

I've been a pediatrician for decades now. I have spent plenty of time in intensive care units and have certainly seen much worse endings. Still, I tear up replaying these scenes in my mind. This is not a medical school application essay. Gabe's hospitalization was no moment of career epiphany for me. But there was a clear message in that experience, and it is one that I consider just about every day. The upshot is this: There is danger in the world, particularly for children—real, present, and meaningful danger. Sometimes we are lucky enough to have figured out ways to mitigate some of that risk. Think seat belts, childproof medication caps, vaccines, and yes, pool fencing. And sometimes when prevention fails, the outcome falls to luck.

The prevention side of things failed for Gabe. Nothing stopped him on his journey to the backyard. He survived because his mother knew what to do, a well-stocked ambulance was near enough, and the magical machines and people of the ICU were available to him in the wealthy country of his birth. He was lucky. Our whole family was. I think about that a lot.

I went to medical school and did my residency at the hospital where they took care of my brother. Working in that pediatric ICU was inspiring and, if I am being honest, mind-bendingly exhausting. I was never cut out to be a critical care doctor, but I'm a puzzle solver, and I

INTRODUCTION

found a subspecialty that lets me spend my days caring for children with complicated diseases and thinking about how to make kids a bit safer tomorrow than they are today. The field of pediatric infectious diseases (ID) is perfect for the way that I think. It's a little slower than critical care. We spend a lot of time digging through patient charts for clues to a difficult diagnosis, consulting journal articles and textbooks, and often calling a colleague who might have the experience or the different approach that we need to turn a stream of contradictory data into a coherent story. As a result, my pediatric ID friends and I also accept a gentle ribbing from other specialists about the lengthy and sometimes absurdly detailed medical histories that we take.

Diagnosis is at the heart of being an infectious diseases specialist, and it was what got me hooked on the field. In front of my medical student eyes, weeks of unexplained fever became a newly confirmed case of tuberculosis that responded quickly to medications. Questions, a physical examination, an X-ray, a lab test or two, and a cure. A different pace and a different kind of magic from what I had seen at Gabe's bedside. During the first week of my infectious diseases fellowship, when I was a fully trained pediatrician but just starting my subspecialty education, I saw a child who had recently returned from visiting relatives in Côte d'Ivoire. He had been fine for the weeks in West Africa and even on the flight home but just a few days later had become extremely ill—high fevers, vomiting, and eventually convulsions. The doctors in the emergency department stopped his seizures with medications. Thanks to the ICU team, he was stabilized and carefully monitored.

My job was to help figure out what caused his dramatic change, and as I entered his room, I silently ticked off possible causes of fever in a returning traveler—typhoid, malaria, pneumonia, and others—reminding myself of the important questions to ask. The boy's mother held his hand, and both parents were patient with me as I had them go over the story of the trip, the return, their son's illness. Where had they stayed? With his mother's family—in a town, not a city. He had spent

INTRODUCTION

his days outside, running around with his cousins. No, he hadn't had any undercooked foods. No unpasteurized milk. No contact with farm animals. Yes, there were mosquitoes around, but they didn't recall him complaining of bites. No, he had never been sick like this before. This was his first time in a hospital since he was born. He was up to date on his vaccines. "In fact," his mother said, "he has never had to take any kind of medication for anything, not even an ear infection." A voice in my head: *Ah! No prophylaxis for the trip.* I asked, but no one had prescribed medicine to prevent malaria during his time in a place where the infection is prevalent—a too-common oversight, especially for children.

A drop of this child's blood on a glass slide revealed the diagnosis. A little while later, in the hospital's laboratory, the senior parasitologist had to help me as I struggled to focus the microscope, but once he did, even my essentially untrained eye couldn't miss the red blood cells teeming with blue-stained malaria parasites—each one a circle with a darker, thicker projection on one side, like a wedding ring. The fevers, the seizures, his parents' fear—all of that was downstream of the drama that we saw playing out under the microscope. Our team prescribed medicines that I had never heard of before that week, drugs with names that sounded like spells to me—quinidine, mefloquine, atovaquone-proguanil. These and a multitude of other images and spells are the knowledge that I now get to teach to a whole new generation of students, residents, and fellows.

Even before we had a global pandemic with which to contend, the past few decades have been an incredible time to be interested in infectious diseases. My training in urban hospitals in the United States focused on caring for individual patients. I strove to be the best doctor that I could for the patients in front of me. Diagnose, treat, repeat. Even with the steady workload, some of the broader trends we saw in the hospital and the clinics were impossible to miss. There were everyday diagnoses, like rotavirus, which could cause diarrhea so voluminous that children would come into the hospital with bone-dry lips and

INTRODUCTION

skin that "tented"—staying in place when gently pinched into a mound rather than snapping back into place. Or like respiratory syncytial virus (a hard-to-spell pathogen of young infants that generally goes by RSV). We learned to recognize the quick, shallow, noisy breaths of a child with RSV, to count the number of breaths in a minute, to listen carefully to their lungs to distinguish a kid with an impending pneumonia from one who was on the mend. These were the illnesses that got turned into shorthand. "There's a new kid with rota in 810. On IV fluids, but still looks a little dry." "Keep an eye on that baby with RSV tonight. She's working hard to breathe." We would even plan for these seasonal diseases in advance, knowing how crowded and chaotic the hospital floors would become once winter hit.

There were other infections that we saw only occasionally but worried about daily. Infants and toddlers arriving in the emergency department with fever would often get a "rule-out sepsis" workup, with the goal of figuring out whether they had bacteria in their blood or, worse, their spinal fluid. Most of those evaluations were negative—false alarms triggered by one or another of the million minor infections of childhood—but the ones that were positive were mostly due to *Streptococcus pneumoniae* bacteria. Those rare "hits" made the workups worth doing, as a child sent home with a missed case of meningitis—when those bacteria travel to the fluid that encases the brain and spinal cord—may be dead by morning. Early diagnosis and treatment with the right antibiotics could head off a tragedy.

The most striking thing that I've seen as a doctor isn't a just-in-time diagnosis or a heroic save of a critically ill kid (even when that kid was my brother). The most striking thing is what happens when we turn the corner from diagnosing and treating a disease to preventing it. This shift makes it so that the saves don't need to happen because the child never even gets sick. There is nothing that the pediatrician in me loves like disease prevention. Those scourges of my residency? Rotavirus is mostly gone in countries lucky enough (read: wealthy enough) to have

INTRODUCTION

reliable access to the vaccine—no more yearly surges of children hospitalized for diarrhea. *Streptococcus pneumoniae*? There's a vaccine for that as well. It's a tougher scientific problem (many different types of the bacteria, each needing its own component of a complicated and expensive vaccine), but enough slow progress has been made that we have been able to significantly decrease the number of kids who need blood cultures, spinal taps, hospitalization, and antibiotics. Today, some pediatric residents go through years of training without ever seeing a child with either of these infections—the same things that filled our nights and our hospital beds not that long ago.

RSV has been a tougher nut to crack. In the early 1960s, a promising vaccine candidate not only failed to protect children against RSV but actually made the disease worse in some of the vaccinated kids. It took decades for scientists to figure out exactly why that initial attempt failed, and longer still before new strategies, including vaccination during pregnancy, have allowed pediatricians and parents to hope again that, as with rotavirus, RSV's days may be numbered. It is incredible to imagine after uncountable sick children, scared parents, nights spent in ICU beds, and rare but world-splitting tragedies, that we could be right on the cusp of preventing this disease as well.

In retrospect, I suppose that this kind of change shouldn't have been quite so disorienting for me. The generations of pediatricians immediately preceding mine saw different childhood infections go from front of mind to back of mind to essentially gone from mind. At the close of my residency, my mother marveled at my never having seen a child with Hib infection (a childhood scourge that has nearly disappeared in the United States since a vaccine against it was licensed in the late 1980s). I have never taken care of a child with polio or diphtheria, either. Thanks, vaccines! Never having seen these diseases is not the same as never having thought about them, though. I've read textbooks, journal articles, and, with increasing frequency as I get older, letters and other personal writings of my predecessors from decades or centuries

INTRODUCTION

ago. Connecting back in that way has been valuable. Because of my direct experience, I can easily conjure the bark and whoop of a child with pertussis or the bent posture and primal distress cry of a baby with meningitis. I have a visceral feel for what an uncontrolled chickenpox infection can do to an immunocompromised child. I have seen these things many more times than anyone could possibly want to. The diseases that we don't see anymore are tougher to visualize, and I rely on descriptions from those who came before me to help me understand.

This exercise is important, because nearly all these diseases are not gone from the world—just gone from the privileged place where I happen to work. Gone for now. Despite our best efforts, polio continues to elude eradication and has even returned to the United States and other countries that had eliminated it decades ago. Diphtheria, congenital rubella—both deadly and eminently vaccine-preventable—are still with us. We teach about these infections, remember them, and continue to vaccinate children against them because they remain threats, even if we rarely—or never—see them in the wealthy countries of the world.

The best example of this phenomenon, the vaccine-preventable disease that still keeps me up at night, is measles. Just a generation or two ago, essentially every child got the measles. It was a rite of passage in childhood, involving a few days off from school, an inconvenience for families. Measles was the butt of jokes in popular culture. In 1969, the popular television show *The Brady Bunch* made light of measles, describing one affected child as having symptoms of "a slight temperature, a lot of dots and a great big smile." In that episode, Maureen McCormick, playing oldest sister Marcia Brady, said, "If you have to get sick, sure can't beat the measles," as the spotted children played a board game together. In *Gone with the Wind*, Scarlett O'Hara's first husband's demise from measles, rather than in combat, is depicted as shameful, unmanly. Even the word *measly*, derived from measles, has been used since the mid-1800s to represent something inconsequential or insufficient.

INTRODUCTION

Measles used to be one of those diagnoses that a pediatrician (or an experienced grandmother) might make from across a room, just from a child's appearance and demeanor. The first symptoms—a fever, red eyes, a runny nose, and a cough—could go along with any of a multitude of childhood maladies. That uncertainty fades after a few days, when the measles rash blossoms—first on the face and neck, then spreading to the trunk and limbs. Raised and flat areas of pink coalesce into large splotches. The rash associated with measles is distinctive enough that the medical adjective that describes it, *morbilliform*, translates to "looks like measles."

The measles vaccine, first available in the United States in the early 1960s, changed this state of affairs. There were fits and starts to its acceptance and widespread use, but within a few decades, measles was rare, and by 2000, it was gone from the United States except for sporadic cases in international travelers. As a result, a generation of parents and pediatricians grew up with measles out of sight and, consequently, out of mind.

In the autumn of 2018, a mother and father brought their young son to one of the hospitals where I work. He'd had a fever, a runny nose, and red eyes for a couple of days, as well as a rash that started that morning. As the rash progressed, his parents headed to the pediatric emergency room, worried that he might be having an allergic reaction to eye drops. He had never needed drops before, but his eyes seemed so red that they assumed that they must be bothering him. Both mother and father were aghast at the pediatrician's suggestion that their son might have measles. "This was nothing. Just a little cold. He reacted to the drops!" Neither parent had ever seen a child with measles, maybe had never even heard of one outside of stories from their own parents. Nor had the pediatrician who first suspected the diagnosis. The knowledge that he relied on came from others' experiences, and his quick realization helped make sure both that the child got the right care and that others in the hospital did not get exposed to a highly contagious

INTRODUCTION

disease. When my team arrived to help confirm the diagnosis, his parents were still looking for someone to say that this was all a mistake, that the tired, sad-looking child with the weepy eyes, the runny nose, and the rash creeping across his face and onto his trunk really did just have a reaction to his eye drops. We couldn't do that, of course. Instead, we sent the testing that would eventually confirm the diagnosis of measles, talked to his parents about what to expect and what to watch out for, and prescribed vitamin A, which can decrease the risk of complications of measles. That child turned out to be an early case in what would become the largest measles outbreak that New York City had seen in a long time. There were hundreds of measles cases over several months, many of whom required hospital admission, as well as millions of dollars in cost for control measures (vaccine drives, school closures, an emergency declaration) and widespread anger in the involved communities.

Many of those children with measles and its complications were admitted to our hospital, and I saw more measles cases over those few months than I had in the prior two decades of my career. At the height of the outbreak, a medical student who had joined our team just a week or two earlier, having never seen a measles rash, started to present a case to the group by saying, "This is another measles kid who was admitted today . . ." I stopped her there, asked her to reflect on what she had just said. "Another measles kid." Measles had gone from being a zebra (an unusual diagnosis, as in the classic admonishment to medical students, "When you hear hoofbeats, think horses, not zebras") back to humdrum in an incomprehensibly short time. Our team went to see that child, who was in the intensive care unit for careful monitoring and respiratory support. She had developed pneumonia and was having significant trouble getting enough oxygen through the inflamed tissue of her lungs and into her blood.

As we approached the room, I was ready for a difficult discussion with her parents. The measles cases in this outbreak were, unsurprisingly,

INTRODUCTION

concentrated among unvaccinated children, and I knew from our student's presentation that this patient's parents had refused vaccines for her, including one that protects against measles. Conversations with vaccine-hesitant parents could go in several different directions. With a hospitalized child, everyone's top priority had to be the current situation—getting her well enough to breathe on her own, leave the ICU, and head toward home. Bringing up the vaccine decision required walking a fine line. How do you explain to a parent that the reason that their child is in the hospital is because of a choice that they made? Carefully. With empathy. But clearly.

Carefully, because asking about this earlier, fateful decision could provoke feelings of defensiveness ("Why are we talking about this now? Shouldn't you be working on getting her better?"), guilt ("Are you accusing me of doing this to her?"), or outright anger. I've certainly seen all three, sometimes from the same family in the same conversation. Over the course of the measles outbreak, I had also encountered parents who took a completely different tack. Not only did they defend their decision to delay vaccines or forgo them entirely for their kids, but they saw the child's eventual recovery as vindication for that course of action, thinking that now there would be "natural" protection instead of the supposedly inferior, artificial, pharma-made kind. In this twisted logic, the outcome proved it. Despite the hospitalization, the blood draws, the fear—measles was no big deal. Doctors have different approaches to this kind of conversation, and some avoid it altogether. I usually lead with curiosity. "We've discussed that your daughter has measles. Can you tell me a little about what you've experienced or heard about measles or the measles vaccine?"

With empathy, because after all, these are the parents of a hospitalized child. Regardless of the decisions, fear, or distrust that brought them to this point, they are in a situation that all parents fear. Innately and completely, we want to protect our children, scare away the monsters, make sure that they make it home safe. Ironically, that is exactly

INTRODUCTION

the response that leads some parents to refuse vaccines. Vaccine-hesitant parents aren't people who want harm to come to their children. They love their children and want to protect them but have generally been influenced by at least two kinds of false messages: fear of vaccines and minimization of the real dangers of the diseases that those vaccines protect against. "Would you be willing to share some of what you heard that led you to the decision not to vaccinate?"

Clearly, because the problem of vaccine hesitancy neither begins nor ends with this particular child in this particular hospital room. The way that this conversation goes can help guide this family's future choices, alter their decision-making about their other kids (or even their decisions about future vaccines for the hospitalized child), and shape the story that they communicate to friends and family about this frightening experience. "We are on the same team here. You want your kids to grow up happy and healthy, and so do I. There are many illnesses we can't prevent, but vaccines protect us against some really serious ones. You didn't allow her to be vaccinated because you were given bad information. We can't go back and fix that, but I want to make sure that we get all your questions answered about this vaccine and any others on the schedule so that we can help keep her healthy going forward."

These can be long and difficult conversations, sometimes playing out over several days—sometimes shut down immediately by families who refuse to even hear the messages that we try to communicate—and their effectiveness in the case of a particular family can be difficult or impossible to determine.

Surprisingly, the conversation in this child's room fell into none of these categories, because her parents were not there when we arrived. Her grandparents were, though, and they were not happy. When I asked about their experiences or knowledge about measles, the emotion that I encountered was anger. They were furious at their own child—the patient's mother—for refusing measles vaccination for her daughter. "We remember measles. We remember polio! We vaccinated our

INTRODUCTION

kids. Why would anyone possibly do this?" Their own experiences had immunized them against the anti-vaccine messages. It wasn't as if they hadn't heard the rumors about vaccines. It's that those messages seemed absurd in light of their experiences. They had seen children suffer and adults who carried the aftereffects of diseases like polio. Vaccines were something to be thankful for, not a dark conspiracy.

What happened between those two generations, the vaccine-hesitant mother and her own parents, dumbstruck at the idea of someone opting out of the miracle of vaccination? How did we go from eliminating measles in the United States and talking seriously—realistically—about its worldwide eradication to a massive resurgence in 2019, presaged by smaller blips in the prior years? And should those blips have warned us about some of the problems that we would face with the COVID-19 pandemic—the vaccine hesitancy, the demands for unproven (and sometimes rigorously and scientifically disproven) medications, the distrust of doctors and public health professionals, the underestimation of that new virus's effect on children?

Before this outbreak, I had seen only a handful of measles cases over the course of my career—an international traveler, a small cluster in northern Manhattan—but I knew that it was an important cause of childhood mortality. Even when we considered it "under control" in wealthy countries, measles remained a major killer in areas where we failed to deliver health care, especially vaccines, effectively.

What I came to appreciate—what I hope to communicate here—is that measles isn't just one more childhood disease. It is the quintessential human pathogen. Measles does more than cause fever and a rash. It teaches us about ourselves—our capacity to learn, remember, and forget. It moves through crowds with astonishing speed. It has shaped and continues to shape our history. We have a vaccine to prevent measles, one that is battle-proven to be able to eliminate the disease from whole countries, even continents. However, what we've accomplished with that vaccine isn't just an example of a scientific triumph, as the

INTRODUCTION

persistence of measles also provides evidence for our ongoing failures to reckon with poverty, inequity, racism, and the myriad legacies of colonialism. Vaccination has decreased global childhood deaths due to measles by an incredible measure, but a disproportionate number of these deaths still occur in poor nations. This pattern of inequity stretches back for centuries, when colonial ships brought the synergistic combination of violence and new infections, including both smallpox and measles, to the Americas and Pacific Island nations, leading to widespread suffering and political turmoil. The more recent history of measles in the United States has still been one of division. U.S. redlining policies determined housing access and financial opportunities based on geography, race, and class. Similarly, measles "rashlining," driven by limited access to vaccination and medical care, gave disadvantaged children worse odds of avoiding the most severe outcomes of the disease. These issues are illuminated by measles but not confined to it, and recent challenges—Ebola, COVID-19, mpox, and others—have repeatedly demonstrated the urgent, persistent, and widening inequities that both precipitate and follow pandemics.

Despite the severity of her disease, that child in the intensive care unit had an outcome typical for children with measles who, by chance of birth, live in high-income countries like the United States—a steady recovery over the course of about a week. Even here, a handful of children out of a hundred with measles require hospitalization for pneumonia, as she did. Less common and more severe complications of measles—brain inflammation or death—happen once or twice per thousand cases. These numbers sound low, but prior to the development of a vaccine, there were millions of cases of childhood measles in the United States every year. That adds up to many thousands of these rare and devastating outcomes, each one a very real child. As we will discuss in subsequent chapters, the outlook for children with measles was significantly worse in the not-too-distant past, even in large cities in the United States and Europe. Even now, as with nearly all

INTRODUCTION

determinants of a healthy childhood, emerging unscathed from measles infection has more to do with the circumstances into which a child is born than with any other factors. A child in a refugee camp in the Democratic Republic of the Congo, which in recent years suffered explosive and overlapping outbreaks of Ebola and measles, is not accorded the same favorable odds as one in our ICU. Malnutrition, crowding, and stress—both physical and psychological—all place a finger on the wrong side of those scales.

Those are some of the reasons that I am obsessed with measles. Despite medical advances, measles is still with us—very much so—and it is a bellwether for our weaknesses, particularly when it comes to public health infrastructure and trust in science. As I write this, the COVID-19 pandemic has waned, yet it persists. Issues of infectious disease epidemiology, vaccine development, rationing of care, and masks have filled news coverage for the past several years. Our global responses to this pandemic have included incredible successes (rapid vaccine development) and marked failures (poor communication about school closures, quarantine, and mask mandates). Trust in medicine and public health is at a nadir. False "experts" use their platforms to sow confusion and antiscientific sentiments. Some of them even run for president. Declining vaccine confidence leads those of us who spend our lives in those fields to brace for not only continued struggles with COVID-19 in the coming years, but also a global reemergence of other vaccine-preventable diseases. Specific biological properties of the measles virus make it an early warning sign of danger to come. Given its unmatched ability to spread from person to person, when it gains access to a susceptible population, measles is explosive.

Every single child diagnosed with measles anywhere in the world represents a system failure—an inexcusable unforced error. The technology to prevent essentially 100 percent of measles cases has been in our hands since well before the moon landing. Every childhood death from measles since that time, and there have been many millions of

INTRODUCTION

them, has happened because we haven't moved fast enough, haven't all pulled in the same direction when we could have.

Measles is a bully. It has always picked on the weak, the undernourished, and the crowded. Well over a century ago, in *How the Other Half Lives*, the journalist Jacob Riis described how measles ravaged the children of New York City tenements:

> Under the most favorable circumstances, an epidemic, which the well-to-do can afford to make light of as a thing to be got over or avoided by reasonable care, is excessively fatal among the children of the poor, by reason of the practical impossibility of isolating the patient in a tenement. The measles, ordinarily a harmless disease, furnishes a familiar example. *Tread it ever so lightly on the avenues, in the tenements it kills right and left.* Such an epidemic ravaged three crowded blocks in Elizabeth Street on the heels of the grippe last winter, and, when it had spent its fury, the death-maps in the Bureau of Vital Statistics looked as if a black hand had been laid across those blocks, over-shadowing in part the contiguous tenements in Mott Street, and with the thumb covering a particularly packed settlement of half a dozen houses in Mulberry Street. The track of the epidemic through these teeming barracks was as clearly defined as the track of a tornado through a forest district. [emphasis mine]

This description rings true today. Wherever in the world there is war, famine, or disaster, along with the refugee camps, you can find measles. In Riis's New York, vaccination against measles was still far in the future. We no longer have that excuse. We have shown that when we are operating at our best, international cooperation can use vaccines to drive diseases from entire countries and, in at least two cases (one of which was measles's closest relative), eliminate them from the planet. Measles thrives on our failing to use that know-how. When we forget

INTRODUCTION

to pay attention, it rises again. Even in wealthy nations, when anti-vaccine charlatans and pseudoscience peddlers thrive, when funding to vaccination programs is cut, when well-meaning parents do not learn how to tell reliable information from its opposite and thus fail to vaccinate their children, measles is often the first sign. It is also a sure indication that other problems are not far behind. These chapters explain why measles is that early alert, why it has reemerged at this point in history, and how we might have heeded the warning that the 2018–2019 measles surge gave us to better prepare ourselves for the approaching coronavirus pandemic and the additional challenges that would follow.

The very weaknesses that COVID-19 has exploited were made clear by the resurgence of measles in both high- and low-income countries in the years leading up to the pandemic. Our history with measles should have taught us that developing a vaccine is necessary but certainly not sufficient for control of a rapidly spreading virus, that politics and public policy need trusted scientific voices representative of the communities they are addressing, and that effective messaging is both difficult and critical. Logistics—the way that infections, public health interventions, and vaccine information intersect with the day-to-day lives of people—is among the most important factors of all. Community members need to understand what those official voices are saying, sure, but their advice also has to fit with the rest of people's lives. It has to be clear, actionable, and feasible—areas in which we have frequently fallen short.

Measles is a biological agent that preys on human inequity, thriving under conditions of chaos, colonialism, and war. It is both a cause and a direct beneficiary of our propensity to ignore and even forget threats that no longer seem imminent. The virus that causes measles is a real, physical entity that interacts with our real, physical cells. Too small to see, it rides droplets through the air, sneaks in with our breath, and exploits uniquely human weaknesses—pushing our immunologic buttons in a way that only a longtime nemesis could. That interplay, between a virus and us, is where our story begins.

1.

Meet the Measles

Is there anything more consequential than a child's breath? In some ways, it is the most routine thing in the world. Tens of thousands in a day—over and over and over—but some breaths are more consequential than others. First breaths can be particularly fraught. My daughter was born when I was a pediatric resident, at the same hospital where I worked. Most mornings, I entered the building in an emotional state that lived somewhere on the spectrum between interest in what I might learn and dread at the mammoth list of tasks that I would have to complete before I could leave again. When my wife Shari and I walked in together, headed for the labor floor, I had a different kind of anticipation—a combination of excitement at the upcoming new chapter of our lives and abject terror at everything that could go wrong. I attended many deliveries as a resident, most of which were joyous, some of which were both tragic and, especially at that moment, easy to remember. Despite efforts both conscious and unconscious, you retain what you see as a physician, and as is true with a duckling imprinting on its parent, sometimes what you encounter earliest in training has greater clarity, more staying power. Uterine ruptures, stillbirths, amniotic fluid embolisms—I had seen enough go wrong in delivery rooms that I was more frightened than was reasonable or logical. Samantha was born with her umbilical cord looped once around her neck—just once, not tightly, not a dangerous situation. She was blue when she first emerged, and I remember waiting for her first breath, repeating quietly the phrase I used when I was the one bringing a newborn patient to the warmer for assessment—equal parts greeting and prayer, "good baby, good baby, good baby." Sam's breath came soon, and with it a pinking of her skin, a fidgeting of her limbs, and the beginning of a life. As any parent who has watched over a child knows, breaths can be filled with meaning.

Our bodies take breathing pretty seriously, but the mechanics of the system are surprisingly straightforward. The oblong muscle of the diaphragm—a crosswise border at the bottom of the rib cage—contracts and descends, increasing the space available to the organs of the thorax (the heart, the lungs). More space in the lungs means a drop in pressure, which causes air to flow in through a system of tubes—starting at the nose and the mouth, branching repeatedly along the way, and ending deep within the lungs. The inner portion of those tubes—where the liquid would be if they were straws—is the lumen, a path containing the air that we have just inhaled. Prior to the breath, that air had been most definitively outside of us, but now it inhabits a liminal space, both outside and in. The diaphragm relaxes back up, and the pressure difference is reversed—the air leaving our lungs, back out the way it came. Over and over, every single day. Our bodies have to maintain this cycle if we are to survive. Our airways, branch by branch by branch, represent a deep and ever-present connection between our human selves and that which surrounds us but is separate from us.

This arrangement, with branching tubes inside of us containing the air that we inspire, implies that there are locations where inside and outside meet. There must be a luminal border, a place where the last point made of human cells abuts the first point of the outside world. That border prevents some things from crossing into the tissues beneath—the protected, sterile parts of our lungs—and allows other things through. We don't move air in and out all day every day for no reason. In the lungs, oxygen crosses the border into our blood, destined to be bound by the hemoglobin packed in our red blood cells and delivered throughout the body; carbon dioxide travels the opposite path, across the border and back into the airways, ready to be exhaled. It is a system set up for constant, regulated exchange.

We have cells maintaining that important boundary between inside and outside. The epithelial cells that form the border crowd together tightly, making it difficult for anything that is not supposed to get through to do so. And other cells act as defenders of that barrier as well. Macrophages reside in the air spaces, where they ingest and destroy debris or microbes, or alternatively process what they have found and present it to other immune cells, which gather in nearby lymph nodes. Dendritic cells live in the tissues below the border, but they extend tiny projections into the lumen—fingers in a grab bag—sampling its contents, evaluating potential threats. It's a pretty impressive defense system, multilayered with built-in redundancy. Much of the non-air matter that we inhale is filtered out high up in the airways—trapped by mucus, ejected by the never-ending beating of whiplike cilia projecting from cells, pushing things up and out. What makes it into the lower airways is evaluated by these sentries—the macrophages, the dendritic cells—which can deal with threats or call in additional immune cells to help. It is striking, then, how effectively measles virus subverts this system.

Like a first one, other breaths can be important events in the life of a child, even if their significance is easy to miss at the time. The first contact between a child and the measles virus isn't usually dramatic or even noticeable. Measles virions (viral particles), suspended in airborne droplets of saliva, mucus, and cells, enter silently, riding a breath. Some of those droplets follow the path of the tubes deep into the lungs, where they encounter the epithelial barrier and its protectors. There, the measles virus commits its first act of deception against its human host, but certainly not its last.

The outer surface of a virion is speckled with proteins and sugars that can fit snugly into receptors on a host cell—a key sliding into a lock. Once that connection is made, the cell itself may pull the virus inside, unaware of the coming havoc. These receptors often play a role in normal cellular functions like acquiring nutrients or gathering information about the cells' surroundings. Diabolically, viruses repurpose the machinery of ordinary business to enter the inner part of the cell, an essential first step toward turning that cell into a mass producer of new virions. H protein, located on the surface of a measles virion, binds specifically to a human protein called SLAM. This interaction triggers a subtle rearrangement of other viral proteins, allowing the virion's membrane to fuse seamlessly with that of the human cell. In a moment, virus and host become one.

Microscopic interactions like this one mean everything for a virus and determine how it manifests in our macroscopic world. That lock-and-key requirement for specific receptors can determine viral host restriction (Does a particular virus infect animal, plant, or bacterial cells? What kind of animal? A primate? All primates or just humans?) and cell tropism (All human cells or just human lung cells?). Viruses can have strict host restriction, meaning that they infect only a single species or a small set of closely related species, or they can be more promiscuous. Influenza wouldn't be the ever-changing worldwide threat that it is without its ability to infect multiple species—its receptor binding more foot in sock than key in lock. The life cycle of the influenza virus, which includes generation of genetic diversity during infection of birds and pigs, coupled with a prodigious ability to infect humans by targeting cells in both the nose and the lungs, make it a perennial and at times existential threat.

Like influenza, the SARS-CoV-2 virus that causes COVID-19 can infect multiple host species. It most likely first emerged in bats but was able to jump (or "spill over") to humans because our cells happen to have a protein similar to the receptor that the virus uses to get into bat cells. This factor is absolutely crucial. We have endured a multi-year global pandemic on the basis of these particular proteins fitting together. And it's not just the fit—the location of receptors matters. Because that human receptor to which SARS-CoV-2 binds happens to be expressed in our airways, the virus can enter and cause damage there, allowing it to spread through our coughs and to infect a new host when they inhale. Likewise, human immunodeficiency virus (HIV), with a host range restricted to humans, targets specific cells of the immune system with the surface receptors that it needs for entry. The resulting immunodeficiency that accounts for HIV's lethality is a by-product of its propensity to use receptors on immune cells. The disease is in the details.

Under normal conditions, SLAM—the measles virus receptor—helps immune system cells interact and exchange information. Macrophages and dendritic cells have lots of SLAM on their surface, but the epithelial border cells do not, so measles virions infect the airway sentries, a counterintuitive yet ultimately brilliant plan. Membranes fuse, and the infected cells ferry the virus past the border and into the nearby lymph nodes, where many more SLAM-expressing immune cells live—a feast for those viruses that just Trojan-horsed their way in. In the coming days, newly infected cells will emerge from the lymph nodes, allowing the virus access to the bloodstream—a flurry of invisible activity. From the outside, though, not even a sniffle.

About ten days later, the prodromal phase, with its mild symptoms—a runny nose, a cough, fever, a little redness in the eyes—begins, worsening over the subsequent few days. These symptoms could easily be ascribed to any one of a million other things—a seasonal cold, the flu, even allergies if the fever isn't too prominent. The child might still be well enough to be in school. One clue can reveal the diagnosis at this point, but it is often missed. Small red dots with a bluish-white hue at their center appear on the inside of the cheeks. These are Koplik spots, named for the New York City pediatrician who wrote about them in the late 1800s—the first to realize their specificity as a harbinger for measles.

By the time the telltale rash emerges—first on the child's head and neck, and then spreading to the trunk and finally to the arms and legs—a good two weeks have passed, long enough that the details may have grown fuzzy, making it challenging to trace back when the original exposure occurred. The measles virus has made billions of copies of itself and has spread to nearly every corner of the child's body, including a return to the airway, this time entering the border cells from their underside using a different receptor called nectin-4. Some of the prodromal symptoms come from measles infecting and ultimately killing these border cells—the ones that it had previously bypassed. More important, virions and even clusters of whole infected cells enter the airway and escape with exhalation, carried away on aerosols and droplets with every breath, every cough. These droplets carry everything needed for the cycle to start again in a new host. All they need is a breath.

These intricate microscopic interactions between measles virions and our cells affect how we experience measles as individuals and as populations. Because measles binds SLAM, it infects a specific set of cells, evades a border, and gains access to our bloodstream. The interaction between measles H protein and human SLAM is highly specific. The SLAM on the surface of dog cells or cat cells won't work as a measles receptor, which accounts for the virus's narrow host range. That one very specific match between H protein and human SLAM is the main reason that the measles virus can infect only humans—if the virus can't get in, the infection can't even get started. As a result, if you have measles, your pets are safe, but your roommates may not be. Because measles binds nectin-4 on the epithelial underbelly later in infection, when the blood is full of virus-infected cells, it reenters the airway in enormous numbers, ending up in droplets that are particularly well suited for efficient airborne spread.

The spread of measles is extremely difficult to control in a population. The long lag from exposure to the onset of symptoms and a contagious period that begins days before the rash appears amplify that challenge. If a schoolteacher notices a measles rash on a student on a Friday, the virus has likely been spreading within the classroom since at least Tuesday. And spreading is what measles does best. Unless the kids in that class are protected by vaccination or have already had the measles, it is likely that more than 90 percent of them will be infected—twenty-seven or more out of a class of thirty children.

There is a formerly obscure epidemiological concept, R_0, that became widely known during the COVID-19 pandemic. The effective reproduction number, or R, quantifies the average number of people infected by a contagious person with a particular disease. R_0 ("R-naught"), or the basic reproduction number, is that quantity at time zero, meaning the point at which everyone in the population is assumed to be susceptible. Influenza, a formidable infectious agent, generally has an R_0 of about 1 or 2, with variability depending on strain and season. During the 2014 outbreak in West Africa, the R_0 of the Ebola virus was estimated at approximately 2. The SARS-CoV-2 strain that caused the initial waves of infection in late 2019 and early 2020 had an R_0 value of about 2.5; estimates for later variants including delta, omicron, and others were considerably higher. Poliovirus is extraordinarily transmissible, with an R_0 of 5 to 7, about the same as smallpox, but measles is the king, with an R_0 somewhere in the range of 12 to 18. In other words, a child with measles can spread the disease to a dozen or more other children, and much of that spread can occur before the rash appears. When the omicron variant of SARS-CoV-2 emerged in 2021, journalists and public health officials grappling with its marked transmissibility called it "almost as contagious as measles."

R_0 represents a snapshot in time—the hypothetical, everyone-in-the-population-is-susceptible time zero. This is useful when thinking about the emergence of a new disease—think COVID-19 in late 2019, or the appearance of a brand-new strain of influenza. It's less reliable when considering a population that may be partially immune from prior exposures or, if we're lucky enough to have one available, from vaccination. The value of R changes over the course of an outbreak, as some of the afflicted recover and others die and still others are vaccinated or act to change their exposure risk.

Measles's high R_0 and its mastery of the air—entering our bodies when we breathe—means that it moves rapidly through susceptible populations, spreading until it runs out of new hosts to infect. It stops only when it hits a wall of immunity (everyone has either been vaccinated or been previously infected) or of absence (there is no one left to infect in this particular population). For this reason, measles thrives in crowded conditions, when people are unable to avoid breathing one another's air.

Measles is an extraordinary human pathogen, a hunter of people, but it hasn't always been that way. People were around for a long time before measles arrived. Like SARS-CoV-2, Ebola, and a multitude of other viruses, measles reached us by way of a spillover event from animals, likely originating from domesticated cattle around the sixth century BCE. The more we understand about infectious diseases, particularly about when and how pathogens have emerged in the past, the better able we may be to predict where and when new ones might arise. Measles became not only a success story but the most widespread of human pathogens—a virus that went everywhere we did, causing untold damage over the millennia. There are lessons worth learning in the history of measles, and those lessons have practical value in today's world. As the COVID-19 pandemic made abundantly clear, spillover events are not confined to the past, and they can still lead to global chaos.

2.

Storytelling

When I made the decision to become a pediatrician, more than one of my medical school professors asked, "Are you ready to be sick for the next couple of years?" The idea behind their half-joking question is that pediatric residents are so closely exposed to children with infections that they spend the first few years of their training getting ill from everything that their patients have. Of course, we pediatricians claim that the reward at the end of this immunological gauntlet is superhuman immunity for the rest of our careers. Neither of these thoughts is completely true, but neither is totally wrong, either.

On a particularly busy call night in the winter of my first year of residency, I had to deal with an issue that did not appear on my highly detailed, color-coded to-do list. I had arrived at the hospital feeling fine but by midnight was having horrible, profuse, watery diarrhea. I started drinking water to try to keep up with the fluid that I was losing, but I felt myself getting dehydrated—fuzzy, headachy, a little dizzy when I stood. I went down to the pediatric emergency room, where a classmate was the on-call resident. He put an IV in my arm, tanked me up with a liter of saline solution, and sent me back upstairs—better but still not great—to finish my shift. This story isn't meant to glorify the idea of working while sick—for ourselves and everyone around us, we should go home and rest when we don't feel well—but it was certainly the culture to "suck it up" and keep caring for our patients at the time.

At the next morning's rounds, my senior resident commented that

I didn't look great. I told her about my trip to the ER and offhandedly said, "Apparently I ate Oscar's poop." Oscar was an adorable one-year-old who had been admitted to the hospital a couple of days earlier with symptoms just like mine. The reason he was in a hospital bed and I wasn't is that when young kids get dehydrated, they may not be able to take enough liquids by mouth. As a result, they can get dangerously sick or even die. Oscar came into the hospital and got tanked up with fluids just as I did, and once he was feeling better, he started to drink again. A sample of his stool tested positive for rotavirus, which gave us a satisfactory explanation for his illness.

We know an enormous amount about how some viruses work—how they bind and enter our cells, reproduce, and escape our bodies to seek their next host. As a resident, I knew that rotavirus spreads through the fecal-oral route. It replicates in the cells of our small intestine, and billions of virions are shed in the stool. They enter a new host through its mouth—often carried there on a contaminated item (a toy, a stethoscope, a hand)—and start the cycle again. Oscar had crawled all over me when I was examining him and talking to his parents. We take measures to try to decrease this kind of spread—gowns, careful handwashing, disposable stethoscopes in the rooms of kids with diarrhea—but somewhere in that interaction, the virus that he was producing made its way to its next victim: me.

We haven't always understood infections in this kind of detail. Over the course of a few hundred years, people have gone from having essentially no idea how diseases might be transmitted from one person to another, to possessing detailed maps—sometimes at the level of single molecules or even smaller—of how microbes enter our bodies and survive, evade, deceive, replicate, and thrive within us and among us. Maps like those are now key to how we think about infections, design new strategies to prevent and treat them, and address new threats. Understanding whether a microbe spreads through the air like measles or through contaminated surfaces like rotavirus is key to the measures that we use to try to control it. Vaccines and drugs are easier to develop and test when we understand the microscopic interactions underlying infection.

The data-driven approach to infectious diseases was in vogue even before the COVID-19 pandemic began, but that global catastrophe threw it into an even higher gear. We had end-to-end genome sequences of the novel coronavirus within days of its detection in respiratory samples from people in Wuhan, China, allowing vaccine research to begin well before most of the world understood the gravity of the unfolding situation. Researchers designed medicines and vaccines in weeks or months, evaluated them in newly generated laboratory models, and tested them with new, highly efficient clinical trial designs that didn't exist even a decade prior. That same sequence information allowed epidemiologists to track the virus that causes COVID-19 throughout the world, to understand its history and its evolution as it moved through the population. Online dashboards flashed global case counts, and databases filled with genetic data. Laboratory scientists probed SARS-CoV-2, uncovering how it bound to, entered, and subverted cells and how it behaved in experimental animals, giving us an unprecedentedly deep understanding of a new virus in seemingly no time at all. The data fire hose addicted some, exhausted others. Many of us alternated between the two. It was possible to marvel at how much we knew and simultaneously despair at the situation's uncertainty.

Even apart from the urgencies of a pandemic, medicine has progressed from simply describing what is visible—how a person with a particular ailment looks, feels, smells—to generating reams of information about every patient. We measure chemicals in blood, generate detailed images of tissues buried deep within the body, decode genetic signatures of both microbes and hosts. Doctors measure and create things our predecessors could not have imagined, allowing us to categorize disease and target treatments with ever finer precision. Still, there are limits, and when faced with something new—a person whose illness doesn't fit the established narrative, a story that doesn't follow the rules—we do what we have always done. We describe.

Our earliest knowledge about COVID-19 arose the old way—from descriptions. In late December 2019, two physicians, Drs. Zhang Jixian, a pulmonologist at Hubei Provincial Hospital of Integrated Chinese and Western Medicine, and Li Wenliang, an ophthalmologist at Wuhan Central Hospital, raised the first alarms about clusters of pneumonia cases that seemed eerily similar to those seen during the 2002–2004 SARS outbreak. Dr. Li, whose leaked text messages were censored and condemned by the Chinese government, died of COVID-19 in February 2020.

For diseases like measles that we've known about for centuries, our knowledge originates from people describing what they saw, working to categorize illness, to puzzle out how ailments might move from person to person. Within medicine, we have traditions about communicating this kind of information. Some of these date to times when all that a doctor could do was observe, empathize with a patient, or use one of a handful of available potential interventions. By writing down symptoms and signs and changes over time, doctors could not only track what happened to an individual patient but could lay a foundation for understanding disease in a more generalizable way. That kind of communication, writing down what we see and presenting the information to others—mostly to our colleagues—helps both the presenter and the recipient. For the former, it organizes their thoughts. To tell a story, you have to formulate a story, pulling together seemingly disparate pieces of information, deciding what you will include and what isn't relevant. And by doing that work, by paring a narrative down to its essence, the teller sometimes sees a path to a difficult diagnosis or recalls a question that should have been asked the first time around. The recipients of the information benefit as well, learning beyond the bounds of what they might see in their individual practice,

STORYTELLING

strengthening the muscles of processing and synthesizing information. A patient's story might allow us to understand another doctor's thought process, how that person organizes information. Today we have formalized many of these rituals, though they're rarely as dramatic as they appear on television shows, with a frightened junior doctor presenting a case to stern-looking elders.

We might write about a particularly vexing or unusual case presentation in a medical journal, though these kinds of case reports are less common than they used to be. Many journals have done away with them entirely, questioning the value of describing one or a handful of patients outside the confines of a scientifically designed study.

Our daily work in a children's hospital contains ritualized versions of case presentations. On work rounds, students, residents, and attending physicians discuss the minute details of stories like these, one patient at a time, walking around to each room, discussing the logistics of when an X-ray will be obtained, when blood might be drawn, what testing or information or response to medication stands between patients and, ideally, their discharge home. Because we are pediatricians, other pressing issues often arise on rounds—an upcoming birthday, the name of a favorite pet, a predilection for a particular cartoon character—crucial data that may allow us to generate a smile or a moment of connection with the patient when we enter the room.

Even the format of work rounds constantly evolves. The increased mass and speed of information is true for pediatric patients as well as adults. Laboratory testing and imaging and medication lists have all expanded in recent years, and many hospitals are experimenting with scripts and checklists on rounds to ensure that we communicate essential information. We now welcome family members to join our rounds when we discuss their child. This breaks down barriers between doctors, nurses, and families—and it's also good medicine. The people who know and love this child most can make sure that we are getting the details of the history right, that the medication list in our computer system actually reflects what they are giving at home. Parents can be our partners in trying to stave off hospital-acquired infections and can tell us if there are parts of our discharge plan that won't work for them. It's a better system than the one that I trained in, one of cloistered huddles of doctors and only sporadic interactions with anyone else. This new design helps ensure that families understand their child's progress and that we—the team of doctors, nurses, pharmacists, social workers—get a broader perspective on how a child may be feeling or what the family needs.

There are formal, ritualized discussions, and then there are those that are less so. If you put two doctors in a room, it rarely takes long before they start telling patient stories. We all carry a collection of these, some with twisting courses leading to an unexpected diagnosis (all the better if the narrator is the one who put the pieces together). In pediatrics, the stories are often of particular children or families, with tragedy, comedy, or a combination of the two that resonates within us. Some lack a clear resolution, ending with a patient's inexplicable improvement or worsening or stagnation. Was the outcome because of our efforts? In spite of them? Those stay with us—tales from last week or last year or last decade that still provide an occasional itch to scratch, driving us to one more quick database query, another peek at an old textbook. An answer may never appear, and yet we retell the stories to ourselves and others. A little dose of vicarious frustration or wonder for the listener, and a tiny unburdening for the teller. When a case troubles me, I want—I need—to share it with my colleagues.

As the early surge of COVID-19 cases came to New York City in the spring of 2020, I spent a lot of time talking to colleagues about the children and families I saw. The earliest ones were some of the most dramatic. A pregnant teenager with pneumonia in both lungs who got better and then much worse—gasping for air, needing the help of a

STORYTELLING

ventilator. A middle schooler in a cramped and too-warm isolation room, able to get enough oxygen only once the ICU team sat her straight up in the bed, her mother—also coughing, audibly wheezing—at her side. We had a lot to figure out then—not just the puzzle of how to treat the individual patient in front of us but the construction of protocols for getting children with suspected COVID-19 evaluated, admitted, treated—all while keeping them, their families, and the hospital staff as safe as we could. We debated which medications might work, how to address the scarcity of tests, and what criteria might be used to allocate a limited number of ventilators, if it came to that. Each of those was a component of the work, the assembly of plans that took place on videoconferences, at patient bedsides, in hallways—six feet apart and N95-masked.

Even as data accumulated, we told stories and gained knowledge the way that doctors always have, measured in individual patients, then in small patterns. Some colleagues and I wrote a short paper about a few very young children with SARS-CoV-2 infection who had come to the hospital because of fever, with no respiratory symptoms at all. It seems almost quaint in retrospect that this came as a shock to us then, but those unexpectedly positive tests represented a moment of dire realization for us—that the virus causing COVID-19 probably lurked in the airways of many of our patients, not just in the ones with hacking coughs and frightening-looking X-rays. Those few stories changed how we thought about isolation and testing. It broadened our view of the unfolding pandemic, not for the last time. In this way, the COVID-19 pandemic was very different from the measles outbreak that we had experienced a few years earlier. Then, while many of the doctors who cared for affected children had rarely, if ever, seen measles before, there were centuries of medical experience with the disease, an accumulated knowledge that began with physicians tending to patients and writing down what they saw. One young doctor's journey in the mid-1800s was a turning point in our understanding of measles. During a time of political and scientific upheaval, he took a fresh look at a common disease. In so doing, he crystallized practical knowledge for his fellow physicians and, in the process, helped give birth to the new science of epidemiology. Our understanding of measles as a pathogen and as a beacon for other troubles continues with him.

3.

Panum's Map

The Faroe Islands, a remote archipelago north of Scotland at about the midpoint between Norway and Iceland, are volcano-born, interlaced with rapid-flowing fjords. Their basalt layers formed from cooling lava tens of millions of years ago, and their history is one of isolation and austerity. In the 1840s the Faroes hit a sociopolitical inflection point. Since a time far beyond when any living islander could remember, the islands had been under Danish colonial rule. Many Faroese were not thrilled with this situation. Over time, they had ceded control of trade, their written language, and even the curricula in their children's schools. In the early 1840s, the Faroese brought grievances to the national assembly, calling for reforms and even an independent legislature for the islands. In this tense environment, the Faroes developed a problem that required help from the Danish government. In 1846, for the first time in sixty-five years, measles returned to their rocky, forbidding shores.

The Faroes were more than a week from Copenhagen by boat, and Thorshavn ("Thor's Harbor"), the Faroese capital, was both the sole authorized port for commerce and the point of entry for nearly all visitors. In late March 1846, Christen Severin Holm, a widowed woodworker, returned to Thorshavn on the schooner *Havfruen* ("The Mermaid"). Several days later, he called for the provincial surgeon (*landskirurg*). Carl Regenburg, a Danish physician on a temporary posting to Thorshavn, found Holm weak and profusely sweating, with fever, headache, and pain in his joints. Regenburg did not notice a rash, and Holm did not mention that he had recently visited a friend in Copenhagen. Holm's friend became ill within a day or two of their reunion—hospitalized and diagnosed with measles. Regenburg examined Holm but did not know about the exposure in Copenhagen. In his estimation, this was a clear case of rheumatic fever and Holm needed to stay at home and rest in bed. Unfortunately, Regenburg was wrong. The correct diagnosis was measles.

The Faroes were then, and remain, sparsely populated. In 1846, fewer than 8,000 people lived on the islands. A contemporary account describes Thorshavn as containing "about 100 houses, most of them mere huts, stuck in among the rocks without any regularity." Those homes were situated on and around a peninsula jutting into a harbor, overseen by a fort without a cannon. Many Faroese lived on even more isolated farms or in small, loosely associated villages. Everyone on the islands knew the challenges of living there, and the Faroese helped their neighbors in times of trouble. Two men, one a teenager and the other an elderly gentleman, did just that for Holm. They attended to him when his illness was at its worst, returning to their families when they could.

By mid-April, Regenburg realized his error. The elderly man who had helped Holm came to him seeking treatment, and the young man went to see Mr. Nolsø, the only other physician on the islands. The two doctors conferred and agreed on two things—that both helpful neighbors had developed the classic rash of measles, and that the Faroes were in serious trouble. Because no cases of measles had occurred on the islands for the prior sixty-five years, nearly everyone was at risk. In addition, because of Regenburg's misdiagnosis, Holm and his caretakers had not isolated from others to prevent the spread of the disease. Over the next few weeks, exposed family members and friends fell ill, and the outbreak expanded beyond Thorshavn to other villages and other islands. Regenburg informed the local Danish government official, who, fearing that measles would soon be out of control on the islands, requested medical aid from Copenhagen.

The Danish crown responded, sending assistance that could lead one to question how seriously they took the request from the Faroes. No team of seasoned physicians boarded a ship from Copenhagen for the Faroes. Instead, Denmark dispatched two inexperienced men in their mid-twenties, recent graduates of the medical program at the University of Copenhagen who still lacked the full clinical training required for formal licensure (and thus were still referred to as "Candidate"), for the islands. Of the two, Candidate August Manicus was the more logical choice. Claus, his father, had been posted at Thorshavn as the *landskirurg* in the 1820s, so young August, Faroe-born, had spent his early years as a child of the islands.

The second selection—the fateful one—is more puzzling. Like Manicus, Candidate Peter Ludvig Panum had a physician father and was born on an island under Danish rule. In his case, it was Bornholm, in the Baltic Sea—the opposite direction from Copenhagen to the Faroes. Peter was still a teenager when his father died, straining the family's finances. To support himself, Peter tutored younger pupils and wrote small textbooks that he sold to other students. According to several reports, he was an excellent student and passed the qualifying examination that allowed him to begin hospital work, but little in Panum's history indicates what made him emerge as the medical board's choice to stamp out measles in the Faroes. Now, though, these two men were headed to a remote Danish outpost, sent by their government as some combination of things: medical detectives? eager helpers? a backhanded insult to the local physicians?

Manicus and Panum arrived in Thorshavn in late June. At a meeting with local officials, they learned that the cases were waning in the immediate area but rampant elsewhere in the Faroes. Scores of residents had fallen ill. Many had died, and supplies were running dangerously low. Despite the unfamiliar terrain, they decided to split up, with

Panum traveling to the north country and Manicus taking responsibility for the large southernmost island (Suderø) and its neighbors.

At this point in the story it would have been easy for the two young doctors to slip into the banal. They could have traveled the islands, provided sympathy and medical care where they were able, waited for the outbreak to die down, and returned to Copenhagen, report in hand, ready to continue their training with the satisfaction of having answered their country's call.

However, Panum was both exceedingly ambitious and a proud member of the medical avant-garde. The world was changing. By the mid-1800s, medical science was poised for upheaval. Political and industrial revolution, centered in America and France, produced a growing belief in the power of individual liberty and scientific progress to improve humanity's fate. A growing movement of scientific medicine germinated, centered in European cities, especially Paris. Young physicians moved away from practice based on the received wisdom of elders, relying instead on conclusions from empiric, laboratory-based investigation. New inventions were everywhere—the stethoscope, tests to detect protein in urine, improved microscopic techniques that let researchers examine organs and cells with better resolution than ever before. If these tools were added to the hard-won body of centuries of medical knowledge, medicine could, in their eyes, become dynamic rather than static—a field of constant improvement and refinement instead of one of stagnation.

Some must have felt as if there were no limit to the new knowledge and treatments that might emerge in the approaching decades. And yet, not everywhere shared equally in these advances, and not even every doctor in the great European metropolises embraced the new scientific ways. For many practitioners outside of the great academic centers, work continued much as it had before. Notably, both groups had few options for treating patients, consisting largely of sympathy, laudanum (a tincture of opium), purgatives, and bloodletting.

That narrow range of treatment choices was important. Despite the rise of scientific medicine, deaths from infections such as measles, tuberculosis, and smallpox (the sole vaccine-preventable disease at that time) remained commonplace. Sporadic cholera outbreaks terrorized cities. Antibiotics would not enter common use for nearly a century, and even handwashing as a method to prevent infection was controversial. The future still concealed fundamental insights about the nature of contagion and tricks to prevent it.

Three hundred years earlier, Girolamo Fracastoro advanced the concept of *seminaria contagiosa*—contagious seeds—as a mechanism by which diseases might move from one person to another. However, this idea—contagionism—remained a minority view well into the mid-nineteenth century. Even around the time of the rise of scientific medicine, belief in filth and crowding as causes of disease was the more mainstream position. In that framework, lifeless emissions arose in swamps, rivers, and crowded tenements. A healthy person who encountered such a vapor ("miasma") could become ill. Thus began cholera outbreaks, consumption, and other maladies. This theory had the convenient side effect of explaining why some diseases disproportionately affected those who lived in crowded, poorly ventilated conditions. It also had the inconvenient quality of being completely wrong.

Six years before Panum's journey, Jacob Henle, one of few contagionism stalwarts at the time, wrote of invisible transmissible living agents as potential origins of disease. Predictably, Henle's arguments were not convincing to many adherents of the miasma theory. Even within his own work, Henle hedged. He called some diseases, such as malaria, purely miasmatic; some a combination of contagious and miasmatic (he placed measles in this group, along with smallpox, and cholera); and a small number of others—those for which transmission could be readily traced to a discrete encounter—purely contagious (syphilis, rabies).

In subsequent years, practical knowledge, observation, and

deduction—the tools of scientific medicine—led more doctors to argue that many illnesses should fall into that "purely contagious" category. In 1843, Oliver Wendell Holmes Sr., a young American physician who had trained at École de Médecine and returned to Harvard as an advocate of the rigorous Parisian approach, turned his attention to a dreaded disease, puerperal fever. Childbirth remained a dangerous prospect in the mid-1800s. In England and Wales, nearly six births out of a thousand ended in the death of the mother. Young women died of overwhelming infection just days after giving birth, and frequently these cases occurred in clusters—no cases for a long period, followed by several in rapid succession, often of women under the care of a single physician. Holmes reviewed the available evidence, considered mechanisms by which the infection might arise, and concluded that physicians and nurses unwittingly carried puerperal fever from one patient to another. In his view, multiple cases of this disease in the patients of a single physician (thus suggesting spread *by* the physician) "should be looked upon not as a misfortune but as a crime." The response from the obstetric community was, as Holmes had predicted, swift and negative, bordering on violent. Miasmas and bad luck were a much more comforting story to doctors who had lost patients to this disease than the notion that such deaths were caused by physicians and, as such, preventable.

Even though further insights would crystalize that the germ theory—Ignaz Semmelweiss (who came to the same conclusions as Holmes about puerperal fever and went even farther, showing that simple hygienic interventions could prevent its transmission) and others who demonstrated the actual living agents of contagion—was still in the future, the miasma hypothesis was faltering by the time Panum and Manicus set out for the Faroes. The two young physicians grew up in this new atmosphere, and when they traveled to the Faroes, they brought not just medical skills, but also deep curiosity and the sense that a great deal of prior knowledge about how disease originated and spread was probably wrong.

In this context, measles was a particular conundrum. It wasn't a new disease—that much was clear. The oldest medical writings that clearly describe measles come from Abu Bakr Muhammad ibn Zakariya al-Razi (frequently written in anglicized form as Rhazes), a Persian physician who lived in the late ninth and early tenth centuries. Rhazes's mindset had quite a bit in common with the "scientific medicine" that would follow centuries later. He believed that careful observation of patients, their symptoms, and their responses to therapies could reveal truths that had eluded others, even the great Hippocrates and Galen. His *Kitab Al-Jadari Wal-Hasba* (*Treatise on Smallpox and Measles*) was consistent with that philosophy. For the first time, he precisely defined and differentiated smallpox and measles, previously thought to be different levels of severity of the same disease.

In the late 1600s, Thomas Sydenham, a man with unorthodox ideas about both politics and medicine, gave European medical audiences their own definitive description of measles. Sydenham believed in the doctrine of sensualism—the concept that meaningful knowledge could be derived only from direct experience and sensation, not theories and academic pursuits. In his mind, medicine was practical work, but doctors needed to be as observant as artists, describing the smallest details as accurately as a painter might represent them in a portrait. In *Observationes Medicae*, Sydenham described the clinical features of measles, demonstrating both remarkable attentiveness and an appreciation of the disease's variability across individuals—from the "shivering and shaking" and "weeping of the eyes" in the prodrome to the "little red spots, just like flea bites" that appear on the head on about the fourth day, increasing in size until they blend together in confluence. Sydenham touched as well as looked, warning his readers that the rash was raised but to a degree "that can scarcely be determined by the eye." The experienced physician can discern this "by feeling the surface with the fingers."

By Panum's and Manicus's time, measles was a common childhood affliction in crowded European cities. Any parent could see that measles attacked groups of children within homes and schools. An exposed pupil might bring the disease home to their younger siblings. Even the senior, learned physicians, steeped in their miasmatic thinking, realized that measles seemed to spread from person to person in these circumstances. Yet there seemed to be variability in how long after exposure measles would occur. Multiple members of a family or students in a classroom might develop measles simultaneously or separated by days or weeks. The disease seemed to erupt from nowhere—and so the miasma hypothesis persisted. Ironically, it was measles's ubiquity in dense populations that prevented a clearer picture from emerging. If there was always someone, or lots of someones, with measles in Copenhagen or Paris or London, incidental exposures were likely to be missed and chains of transmission misunderstood. Clarity, when it came, would require observations from a remote place housing a highly susceptible population that had minimal contact with European metropolises, a place where the right observer could clearly track the initiation and spread of a disease—as discrete as the bite of a rabid dog—and thus dispense with the need for miasma. The Faroe Islands fit that bill.

The Faroes hadn't seen measles since 1781. As dictated by law, commerce occurred only with Denmark, and the comings and goings of ships were carefully recorded. The port of Thorshavn, where the outbreak began, was as busy as a Faroese location could get, but there were plenty of other places on the Faroes that housed only single families or small groups. Some tiny islands had merely a handful of people whose sole outside contacts were occasional visits by boat. The cold, the violent currents, and the sheer cliffs separating the islands kept people from swimming from island to island. Panum noted that a visit from another island was such an important event for many remote Faroese that they would record it in their calendar—a data point for him to pluck.

PANUM'S MAP

Over several months, Panum visited fifty-two Faroese villages, tending the sick, observing the terrain, collecting information from families and church registers. The latter proved to be the best available documentation of births, marriages, and deaths—including causes of death, if they were known. He also knew the size of the Faroese population, down to the individual, from the prior year's census and wrote down who had been in contact with whom.

One of Panum's first major insights came from Tjornevig, a remote hamlet situated in a deep valley at the northwestern corner of the island of Strømø. By the time he arrived there, Panum had already been mulling over Holm's story and what it meant about the incubation period of measles—that is, the delay between exposure to someone with the disease and the development of symptoms. What did it mean that Holm had been exposed in Copenhagen and not fallen ill until after his arrival in Thorshavn, and what could he make of the fact that those who tended to him themselves became ill a couple of weeks later? Measles clearly had an incubation period, but was it fixed or variable?

The story of the pilot whales and the men from Tjornevig was the key to determining the incubation period of measles. In the 1840s, the commerce, as well as the sustenance, of the Faroes depended on fishing. Faroese fishermen caught cod and herring and salted them for export to Denmark. Other types of fish—haddock, codfish, flounder— were mainly for local consumption. However, there was another fishing experience—one that didn't come around very often—that quickly brought the community together and that continues as a controversial part of Faroese culture today. Schools of pilot whales (called *grind* chiefly by the Faroese), sometimes hundreds of them, appeared occasionally in Faroese waters. An account from the 1800s describes the signals used to efficiently communicate such sightings across the islands (a jacket flown from a ship's mast; the joyful cry of "Grindabud!"), followed by crowds of men armed with spears rushing to boats. These grind hauls could last for days on the close quarters of fishing boats and frequently brought together people from distant villages. The hauls were hard work, but the potential rewards were great—grind meat was a staple of the Faroese diet and could be eaten raw, cooked, or wind-dried for later use—and dividing the spoils at the end of the hunt was cause for celebration. When a haul occurred in the middle of the measles outbreak, it represented a unique opportunity for the virus to seed new villages.

On June 4, 1846, ten Tjornevig men participated in a grind haul off the coastal town of Vestmannhavn. Unlike Vestmannhavn, remote Tjornevig had not had measles cases prior to that haul. The ten men had not left their village before the grind sighting, and they returned immediately after distribution of the bounty. Strikingly, all ten of the Tjornevig men developed the measles rash on the same day—June 18, long after their return home—and their families and friends became ill two weeks after that. When Panum arrived at Tjornevig on July 2, almost everyone in the village had the measles—he estimated eighty out of a hundred inhabitants. The exceptions were the ten original boatmen, who were recovering by then, and a handful of people who would fall ill in a third wave that followed the second by another two weeks.

There was a tremendous amount of information packed into this single village's experience over the span of a month, and Panum appreciated its significance. The small group of villagers had a clearly defined and time-limited exposure to measles, returned home feeling well, and then had a synchronous onset of illness. This was followed by another lag and then chaos as the whole village became ill. In both cases the delay was fourteen days from exposure to the beginning of the rash, preceded by a couple of days of runny nose, fever, and eye pain.

Fourteen days from exposure to rash—like clockwork. Measles had a pace, a regularity that had never been appreciated fully. And now Panum had an epidemiologic tool. In addition, the timing meant that the period of contagiousness had to be right around the rash's first appearance and not during the many days after that, when the rash reached its peak and then slowly subsided. Next came the legwork needed to determine whether his hypothesis would hold up to a larger sample size. He visited more villages, asking questions, gathering data. In each area, he tracked down the earliest known case, determined the date that their rash appeared, and then dug further. His questions traveled forward and backward in time, probing potential exposures and the timing of subsequent illness in their contacts. Everywhere he went, his suspicions were confirmed—the incubation period of measles was fourteen days.

When Panum's predictions worked, they seemed miraculous to the Faroese people. He reported: "[On] account of my observations, I acquired the reputation of being able to prophesy." Only one member of farmer Hansen's ten-person household (his daughter) had gotten the measles. The others thought that they had escaped the infection, as the daughter was already improving when Panum arrived. He counted forward fourteen days from the first day of the daughter's rash and told the family to underline the date in their calendar. On his next visit, Panum was greeted with awe, as the remaining nine had all developed the rash on the predicted day.

Panum's colleague Manicus had a similar experience in the southern part of the Faroes. The island of Store Dimon, separated from Suderø by violent waters and ringed by sheer rock faces, provided a breeding site for multitudes of puffins and other seafowl. However, it was less hospitable for humans, with only a single family residing there. Treacherous crossings made trips in either direction rare. A local clergyman visited once each year, needing to be hoisted up the cliffs with ropes, and the family came to the larger villages of Suderø only when absolutely necessary. One of these visits took place during the measles outbreak. A small group from the family rowed across to visit a trading post and returned to Store Dimon the same day. These men all developed the rash of measles fourteen days later, and the remainder of the family fell ill after another fourteen. The near-total isolation of the island and the efficiency of measles's contagion made the chain of transmission overwhelmingly clear.

Like its 1781 counterpart, the 1846 outbreak eventually decreased in intensity and died out. The two candidates had tended to the sick, but as a contemporary medical textbook put it, "[Measles] is never known to be cut short by art, or abridged of its natural career." In short, measles on the Faroes did what measles always does in a susceptible population—it spread until it ran out of new people to infect. As the tempo slowed, Panum and Manicus tended to some other health issues in the population and prepared to return to Copenhagen. They composed separate formal reports for the board of health and readied themselves to return to normal life in Denmark. Reading these reports is remarkable due to their wide scope and slow pace, particularly in comparison to the terse, confined character of twenty-first-century scientific writing.

Panum's report starts on fairly safe ground. He does not begin with the first cases of measles on the Faroes, or even with a description of the disease as it occurs elsewhere. He zooms out—way out—and explains that he has first to detail the physical nature of the islands themselves, as well as the behaviors and beliefs of the people, before he can explain the outbreak. He begins by justifying this approach:

> When a physician is called to work in a place where climatic and dietary conditions are different from those to which he has been accustomed, his first problem is to study the hygienic potentialities which affect the state of health of the inhabitants. It is, in fact, these hygienic conditions which contribute towards the development and frequency of some diseases and the improbability or rarity of others . . . it is, indeed, on these conditions that the geography of disease, the special study of which

subject will soon, perhaps, elevate it to the status of an independent science, is based.

An independent science! That is quite a place for a candidate physician sent to investigate an outbreak in a neglected colony to start, but Panum's broad framing of the problem at hand explains the unusual scope of what he chooses to report.

At the outset, Panum places himself in the role of a naive Dane exploring the Faroes for the first time, and he makes no secret of the fact that he mostly does not like what he sees. The Faroese winds are "exceedingly uncertain and violent," requiring him to dive to the ground to avoid being blown away. This unfamiliar landscape, with its salty air, sparse vegetation, and "melancholy character," seems far inferior to his beloved Denmark. To his ear, even the songs of the birds sound monotonous and sad.

Perhaps inevitably, given his colonial biases, Panum finds the physical aspects of the islands reflected in the character of the inhabitants. In his telling, the Faroese display a lively exterior that hides an introspective nature laced with insincerity. Panum describes them "as hollow as the cliffs of [their] islands." Having firmly established himself as an outsider who does not find the Faroese particularly sympathetic, Panum describes the habits and living conditions on the islands for his continental audience. Some of his expositions are relevant to the health of the Faroese. For example, he supposes that constant exposure to cold, salty air is the reason behind the high rates of chronic bronchitis. At times he is judgmental of the Faroese—linking their inadequate footwear or shirts that fail to cover their bellies to colds and rheumatism. Other observations border on the prurient, such as his discussions of masturbation ("probably not rare" on the islands) and women's undergarments (not worn).

Panum is repulsed by Faroese dining, especially *drujl* (unleavened barley bread), and wind-dried lamb's meat, which he describes as having an unbearable odor and moldy appearance. At times, he demonstrates some insight, though, realizing that just as he cannot stomach eating raw grind with his shipmates, the Faroese find it just as disgusting that Danes eat aged cheese or salt their meat.

Notably, both Panum and Manicus report on widespread poverty in the Faroes. The islanders' existence is described as one of subsistence work and reliance on the charity of neighbors, particularly for those who become ill. In this way, the young physicians were far ahead of their time. Doctors had noticed links between poverty and illness, but medical treatises of the era did not generally explore specific social determinants of health and disease. To that end, Panum described Faroese homes—shoddy single rooms arranged around a peat-smoke-spewing stove—injury-prone occupations, and rare inclination to wash clothing. He listed all of these as potential predisposing factors for ill health.

Given the forbidding landscape, repulsive and unhealthy (in Panum's eyes) food, and widespread poverty, Panum was shocked at what the local census data and church registers revealed. The Faroese enjoyed a *longer* lifespan than their Danish counterparts (approximately forty-five years in the Faroes compared to thirty-six years in Denmark, by his calculations), due in part to decreased child mortality. Panum's surprise at the health of the islanders is palpable in his writing. How could a poorly educated, provincial, and undernourished population live longer and produce more children than that of Denmark proper? Recognizing the dissonance between his preconceived notions of Faroese life and his data, Panum did the courageous thing—the scientific thing. Shifting his perspective, he honored the data and began to generate hypotheses.

What was it in Faroese life that allowed for overall longevity and safer childhoods? Having enumerated the shortcomings of Faroese culture, Panum focused on its potential virtues. Despite the salt, the air was pure. Quotidian life was outdoors, customs were simple, and the pace was calm. He considered these factors but ultimately judged them insufficient. Panum satisfied himself with an answer found in the periodic historical upticks in mortality present in church registers. Outbreaks of infectious diseases were rare on the islands. Panum concluded that this paucity of infections common elsewhere was responsible for the low childhood mortality and extended lifespan but also left the Faroese people vulnerable to periodic disaster. He had just seen firsthand what measles could do to the population after a long absence. Smallpox had last been seen on the islands nearly a century and a half earlier, and both scarlet fever and whooping cough (pertussis) were completely unknown to the residents. All three of these would go on to cause significant outbreaks in the Faroes in the coming decades.

Panum's creative and expansive framing of his assignment made the current measles outbreak even more notable and urgent to understand. He likely did not realize the outbreak's full scope until he synthesized his findings—of 7,782 inhabitants of the islands, approximately 6,000 (more than three-quarters of the total number of Faroese) suffered from measles over the six months that the outbreak raged. Overall mortality on the Faroes over the course of the epidemic was about double its usual rate—a stunning increase.

Over the course of the second—substantially shorter—part of his report, Panum mines his trove of exposure data, clinical descriptions, ship registers, and understanding of the Faroes. In so doing, he formulates an understanding of measles far beyond anything in the existing medical textbooks. These few pages alternately bust myths and solidify hypotheses about measles that had been part of medical teaching for centuries—maybe longer.

Once again, Panum starts on stable, easily navigated ground. Doctors had varying estimates of the measles incubation period. Eight days. Or ten. Or fourteen. Or there wasn't a defined incubation period. In the chaos of Copenhagen and other large cities, it was impossible to tell. On the Faroes, the answer was clear: fourteen days, give or take a day, between exposure and developing the measles rash. The grind catch story and Manicus's observations on Store Dimon made this clear, as did many additional cases across those fifty-two villages. While the two-week rule held with regard to the rash, Panum noted that one source of confusion was that the duration of the prodromal stage—the handful of days of fever, eye pain, and cough preceding the rash—could vary from person to person.

Likewise, most physicians parroted the teaching that measles was most contagious as the rash faded and the skin peeled. The leading theory was that flakes of skin could spread measles to others. Travel data, ships' logs, and single contacts like the grind haul made it clear. The peak of infectivity was right before and right after the rash first appeared. Conventional wisdom, overturned.

Panum was particularly struck by the effect of measles on Faroese of nearly all ages. In his experience, measles was a childhood disease, but here he found that all but the oldest inhabitants of the islands were affected. Measles deaths concentrated in the very young and those between fifty and sixty years of age. Almost no one over age sixty-five—the group who was present when measles last visited the islands—was affected. Previous infection with measles conferred long-lasting protection. All of those cases of repeated measles infections in the medical literature? Probably misdiagnosed.

How infectious was measles? Crazy infectious. Even brief exposures led to efficient spread of the infection. Did quarantine (closing the doors of a house or a village to all outsiders during the outbreak) work? Yes. Several villages tried this approach, and it seemed to provide some protection. Panum estimated that about 1,500 inhabitants were spared measles that way.

In future outbreaks, quarantine of suspected cases and isolation of the sick would serve the Faroes well. In 1875, multiple introductions of measles came from fishermen from the Shetland Islands. Some of the sailors were visibly ill when they arrived at Thorshavn or Vestmannhavn. Those exposures were sufficient—along with the large number of susceptible Faroese who had been born since the events of 1846—to spark another outbreak on the Faroes, eventually involving more than a thousand people across several islands. However, this time, the local government worked with the Faroese people to use quarantine and isolation systematically to contain the spread of disease, allowing some islands to escape measles entirely.

For Panum, clarifying the stages of the illness and carefully counting calendar days might be of some interest to parents or physicians back home and could help him secure a reputation as a doctor with a keen eye and a quick mind. But he had bigger fish to fry. Having built his argument's foundation on incubation periods and the absence of repeat infections, he set his sights on the miasma theory.

Although prior medical writings described measles and some other childhood fevers as likely to be spread via person-to-person contact, the prevailing view locally and in Denmark was that it was a miasmatic-contagious disease. Defined incubation periods traceable through contacts with people, a clear point in the illness that corresponded to infectivity, and the striking effectiveness of quarantine all built to a clear conclusion: measles was not a miasmatic disease, not a miasmatic-contagious disease. It was spread from an infected individual to a susceptible one—full stop.

Panum also undergoes a striking shift in his initially haughty, colonial view of the Faroes. He advocates for increased resources to be sent to the islands, particularly agreeing with Manicus about the need for a physician on poverty-stricken Suderø. In addition, he criticizes the crown's change of policy that had abolished measles and smallpox quarantine of Faroe-bound ships, pointing out that while such a law would be bound to fail in Copenhagen (because of the constant presence of measles), retaining such restrictions could have saved many lives by blunting or preventing the 1846 outbreak. This represents a curious and substantial evolution for a narrator who started out as a hapless Dane unable to tolerate the local food.

Panum's report included an appendix (one that has not generally been published with translations but is filed at the Danish National Archives), a hand-drawn map of the Faroes, with routes of contact delineated and dates of first cases marked for each village that he visited. This map is a rarity in a scientific publication—something instantly and easily interpretable. A single line enters the map from the lower right, the *Havfruen* sailing directly to Thorshavn. From there, individual lines lead out like a tangle of hair—stretching to nearly every other village. In many cases, that initial spread is all that is listed; in others, those villages act as seeds for secondary spread to more isolated areas. Vestmannhavn, home of the grind haul, is one of those secondary sites of spread. Fuglø, the difficult-to-access "bird island," was seeded twice, but neither time directly from Thorshavn. The routes, the dates of contact, and the careful lines all make the pattern of contagion clear.

After the adventure to the Faroes and submission of their official reports, both Panum and Manicus resumed their medical training. They published twin articles on their experiences in a Danish medical journal. Panum's report became an international sensation, though not an overnight one. It was a masterwork of observation, methodical data collection, and synthesis, but despite Panum's skill, it might have been destined for a Copenhagen file drawer. He had bigger plans. In 1847, he traveled to Berlin, where he met another rising star of academic medicine, Rudolf Virchow. Virchow, less than a year younger than Panum, had just started his own medical journal. Impressed with Panum and his work, Virchow offered to publish it in German in the new *Archiv*, revealing Panum's ideas to a massive and sophisticated medical audience.

After returning to serve in the Danish navy for several years (and to apply the knowledge that he gained in the Faroes to stopping a cholera outbreak), Panum leveraged his newfound notoriety into positions with professors across Europe. He never worked on measles again. When he returned to Denmark as a professor of physiology and a zealot for the scientific method, he made contributions—some of which are still discussed in the twenty-first century—across a range of fields, including immunology, physiology, nutrition, and binocular vision and became a driving force behind integrating Scandinavian medicine with its more traditional European counterparts.

As a young trainee, Candidate Peter Ludvig Panum journeyed to a remote set of islands to help control an outbreak. This decision catalyzed a chain of events with profound implications. In retrospect, it is difficult to imagine a more perfect set of circumstances for understanding the nature of infection. The islands themselves became a natural laboratory: their geography and social structure made tracking the movement of people and measles a tractable problem. Panum himself realized his luck in this regard. He attributed his ability to draw broad conclusions from such a "generally familiar," "almost trivial" disease as measles to the "particularly favorable circumstances" in the Faroes.

In fact, the Faroes themselves continue to teach us. Early in the COVID-19 pandemic, the islands' isolation and widespread availability of testing helped spare them from the massive waves of infection that other nations experienced. When COVID-19 eventually did spread there, reports from Faroese experts provided clarity that was unavailable elsewhere, including one of the earliest and most carefully documented reports of spread of the omicron variant of SARS-CoV-2 within gatherings of fully vaccinated people—a warning, mostly unheeded, for other populations of the potential dangers of relaxing restrictions on masking and large gatherings too early and a Panum-like demonstration of the value of careful observations in small groups.

Not only was Panum lucky and the situation right, but medicine was primed to receive and act on exactly this kind of new knowledge. As Panum predicted in the opening lines of his report, epidemiology was coming into its own as a science. John Snow's London "cholera map" investigations leading to the dramatic removal of the Broad Street pump handle took place only a few years after Panum's report. The historian Jim Downs has chronicled numerous reports from British colonies in the mid-nineteenth century—cholera in Jamaica, yellow fever in Cape Verde, and many others—demonstrating that empire and bureaucracy and colonial exploitation all both kindled the spread of deadly infections and fed early epidemiologic theory. In the decades following Panum's report, his work, as well as Snow's cholera map, and investigations from others helped turn the tide of scientific orthodoxy away from climate and miasmas as the causes of epidemic disease and toward an understanding of specific contagion—transmission of disease from person to person. Laboratory investigations, led by Louis Pasteur and Robert Koch, would help deal the final blow to the miasma theory, but descriptions of outbreaks—particularly those that erupted in the setting of poverty, malnutrition, and neglect—were key to understanding how contagious diseases spread and how they might be stopped. We continue to benefit from these lessons today.

4.

Contagion in the Service of Empire

The Faroe Islands' experience with measles contains a valuable lesson—that structural aspects of human society affect whether and where contagious diseases thrive. The converse is also true. Contagion shapes society.

In March of 2020, a teenager came to the emergency department of one of the hospitals where I work. Her past few days had been difficult, with a spiking fever and a relentless sore throat. At night, she shook, unable to get warm. As her coughing accelerated, her chest began to ache, and pain surged low down in her belly. In her eighth month of pregnancy, she knew that these kinds of pains meant that she needed to find a hospital.

COVID tests were hard to come by in those early pandemic days, but we managed to send a swab from her nose to confirm the diagnosis that we both feared and suspected. She was one of the first people with COVID who I had directly cared for, and as I examined her, I tried to notice everything, to start to put together a mental framework for how someone with this strange new disease looked, sounded, felt. There were the kinds of details I'd expected to find—her eyes were bright red, she had a ton of mucus coming out of her nose, the back of her throat appeared raw—pieces of the physical exam that I would tuck away to think about later, but the parts of that first encounter that I remember most vividly were the emotional ones. The most striking thing about her eyes wasn't the redness, but the fear and the aloneness in them. In pediatrics, most of our patients have a parent or another caring grown-up with them, especially in the hospital. Because of the newly established COVID protocols, this patient's mother had initially not been allowed in, and now this teenager was in a nightmarish situation—facing wave after wave of doctors and nurses, who were all gowned, gloved, masked, and scared themselves. This was a pregnant patient with COVID and pneumonia who needed to be in the hospital, but in that moment, she was also a kid who needed her mom.

Over the next few days, she got a little better and then much worse. We treated her with antibiotics because we were worried about the possibility of a bacterial infection atop the COVID, but we had no specific medicines to help her fight the virus itself. After an emergency cesarean section, she needed an ICU bed and a ventilator to help her lungs recover and didn't even have the chance to hold her newborn baby. That beautiful, vigorous, perfectly well child was born into a chaotic world, feeling the first touch of a family member only when her grandmother was finally allowed to pick her up from the hospital lobby a few days later, the two of them racing through a locked-down city to their apartment to wait for her mother's slow recovery. This story happened early in the pandemic, but it was far from unique. COVID shook and sometimes destroyed innumerable children and their families. There were the obvious tragedies—the loss of grandparents, parents, and sometimes even children to the virus itself—but there were also an uncountable number of stories like this one, in which everyone survived, but COVID still left its mark. Some families were unable to stay with chronically ill children in the hospital, at precisely the time when a child needs support most. In practice, even at the height of COVID, at least one parent could still stay in the hospital with a sick child, but there were strict limitations on who could come and go from the hospital, making a hard situation even worse for already stressed parents. Of course COVID changed everything, even for families that avoided the virus entirely in those early months. Lockdown, work from home, remote school—COVID profoundly altered the way that our societies functioned, and it did so in a matter of only weeks or months. Its effects are ongoing, in some cases maybe permanent. The girl described above, born in the early months of the COVID pandemic, has lived her entire life in a different world from the one in which she was conceived. If we ever doubted it before, COVID made it clear that contagious diseases can bend and sometimes break the world around them, but COVID is a relative newcomer to this game. Measles has been shaping and reshaping human societies for centuries.

Measles could succeed only as a human-specific disease—one that didn't require repeated spillover events—once cities and their surrounding areas reached a population size and density that allowed person-to-person transmission to continue indefinitely. This situation is called endemicity. Epidemic diseases appear sporadically; endemic ones ebb and flow, but they do not disappear. William H. McNeill, in his 1976 book, *Plagues and Peoples*, called such infections "civilized diseases." Despite McNeill's conflation of urbanization with civilization, the point remains that even once measles entered human populations, it was still unevenly distributed across the world—concentrated in urban centers and entering smaller populations only sporadically via travel. In cities, measles's initial explosive spread brought widespread suffering and death, but large numbers of people weathered those encounters. The combination of survivors with lifelong immunity and an ongoing supply of newborn or newly immigrated susceptible people kept the chain of contagion going. In time, measles transitioned from an epidemic disease affecting people of all ages to an endemic childhood disease. As the experience in the Faroe Islands illustrates, in areas without ongoing exposure, measles remained a major, though sporadic, threat. Trade, conquest, and kidnapping brought the virus to these places, overcoming populations with the lethal combination of contagion and war. Empires fell; new ones emerged. The biological basis of measles's success as a pathogen—its incomparable efficiency of contagion—eventually led to its establishment as an endemic childhood infection nearly everywhere in the world, but the road to that outcome was long, deadly, and highly unjust. Colonialism, slavery, and war abetted the spread of measles and increased its toll.

Theories abound about the role of infectious diseases in the conquest of the "New World" by Europeans beginning in 1492. The "germs" in the title of Jared Diamond's *Guns, Germs, and Steel* refer to smallpox, measles, influenza, and other illnesses that accompanied Columbus and subsequent colonizers across the Atlantic. Diamond described the decisive effects of technology (guns, steel) and biology (germs) in the conquest of the Americas, arguing that the combination of these factors was the basis of the European colonizers' success. Many subsequent writers have agreed with Diamond's emphasis on the germs, less consistently on the guns and steel.

Diamond leaned heavily on the concept of "virgin soil" outbreaks to explain the outsize role of infections in European conquest of the people of other continents. Due to their geographic separation from Eurasian population centers, indigenous American populations lacked prior exposure to the microorganisms that caused these diseases. With first encounters came massive deadly outbreaks, affecting the residents and not, for the most part, the newly arrived Europeans, many of whom had immunity due to childhood exposures. A smaller number of infections moved in the other direction, from being well established in indigenous American populations to afflicting the newly arrived and previously unexposed Europeans. The most notable of this latter group was syphilis, which entered Europe with returning colonizers (though emerging evidence suggests that *Treponema pallidum*, the biological agent of syphilis, may have made its way to Europe earlier than that). While syphilis was and remains a cause of considerable suffering, it lacked the wide transmissibility and lethality of the pathogens that moved west across the Atlantic.

The virgin soil theory applies equally to the individuals at highest risk—those at the extremes of age. A pregnant woman who has never had the measles cannot pass protective measles-targeting antibodies to a developing fetus, so even the youngest children would be at risk. The elderly could not rely on protection from past exposures, so they, too, would be vulnerable. Timing of infection is also crucial. It is one thing to have individuals, even many individuals, get sick over time. It is something else entirely to have nearly everyone ill at once, with no one to nurse the weak, bring clean water, harvest the food. Confusion and societal destabilization followed, allowing the colonizers, despite their smaller numbers, to subjugate indigenous populations. Multiple different diseases spreading simultaneously or in close succession amplified the chaos.

Much writing about virgin soil outbreaks teems with a kind of immunological determinism—the idea that measles, smallpox, and other diseases were going to catch up with these populations at some point, that their microbial isolation couldn't last forever. In the same vein, some authors suggest that factors beyond just individual exposure were relevant—that, in addition to the indigenous individuals lacking experience with these infections, indigenous *populations* had not had time to adapt to them, which meant that continued contact between a population and measles or smallpox over generations would have selected for people better able to survive the infection. Newly exposed groups would not have had that head start. The usually unstated corollary of both these lines of argument is that the European colonizers are off the moral hook. Sure, they benefited from disease-catalyzed depopulation in the Americas; sure, importation of enslaved people from African nations became that much more important to the colonizers (and thus accelerated) once local labor sources vanished. But it was all, under virgin soil logic, preordained.

While some aspects of virgin soil theory are probably correct, others are incomplete or flat-out wrong. Most important, these diseases did not arrive in isolation. They were carried to Mesoamerica by groups intent on plunder and on toppling indigenous societies. The effects of measles, smallpox, and other diseases were not preordained. Political turmoil, colonial incursion, and a rapid breakdown of societal structures—all courtesy of the intruders—worked in synergy with the direct effects of infection to amplify depopulation. The flawed deterministic reasoning behind virgin soil theory is cloaked in attitudes of condescension and social Darwinism, bordering on eugenics. Even the choice of words—populations are "naive," "virgin," "vulnerable," humans are "soil"—belies colonial attitudes toward indigenous people.

Most scholars believe that the earliest outbreaks after European contact were caused by smallpox, though multiple overlapping outbreaks of different diseases (including measles) likely occurred. Hernán Cortés's forces ignited a smallpox outbreak in Tenochtitlán, the massive island capital of the Triple Alliance (widely known as the Aztec empire), facilitating the city's fall and the empire's collapse. Contagion cleared the way for conquest. This scenario played out in various forms across Mesoamerica after the fall of Tenochtitlán, culminating in the destruction of other indigenous societies. Serial outbreaks of smallpox and measles devastated Mayan civilization. Farther south, the Inca ruler Huayna Capac and much of his court died of smallpox in the 1520s (there is debate about the exact date), plunging the empire into civil war. This division facilitated Francisco Pizarro's capture and execution of Atahualpa, the last Inca emperor, and his subsequent conquest of Peru.

Though there were probably earlier introductions, by the 1530s, measles had definitively joined smallpox as a cause of deadly epidemics across the Americas, seeded from European cities. Crossing the Atlantic by ship was a more difficult voyage for the measles virus than for smallpox, for reasons rooted in viral biology. Individuals with smallpox can be contagious for several weeks, and dried scabs harbor infectious virus for months. In contrast, people with measles have a restricted contagious period, from a few days before the onset of the rash until a few days afterwards. This means that over the course of the voyage from Europe to the Caribbean, which could last four weeks or more, measles would have to infect serially a chain of susceptible passengers in order for someone to still be contagious when the ship reached its destination. Given that most European adults would have had the measles in childhood, there was no guarantee, even if measles virus managed to board a ship hidden inside a passenger's cells, that it would find enough additional hosts to infect along the way.

Enslaved people from West Africa represented a source of susceptible individuals to maintain a chain of infection during transatlantic voyages. In response to the decline of indigenous populations due to ongoing epidemics, during the 1530s Spanish crews captured and transported people from West Africa to the Caribbean in increasing numbers. Measles was endemic in population centers of North Africa in the 1500s, but many other areas of the continent had not yet reached the necessary population density to maintain transmission, meaning that young people from those regions were less likely to have measles immunity from prior exposure. These stolen people likely served as additional susceptible hosts for the measles virus to cross the Atlantic—linking contagion, colonialism, and slavery in one horrific package.

Following smallpox's path, measles swept through Caribbean islands, across the Mexican coast, into the interior valley, as far south as the Andes, and at least as far north as Florida. Unsurprisingly, indigenous people died at far higher rates than the European colonizers, and by the mid-1500s, waves of smallpox, measles, and influenza brought massive depopulation and societal collapse. Noble David Cook, referring to the combined effect of these infections, wrote: "The foreign pathogens were active, winnowing the people more quickly even than the sword or the arquebus, and certainly much more silently and effectively." The central Valley of Mexico had an estimated population of 25 million people in 1518, just prior to Cortés's arrival. A century later, following wave after wave of imported contagion, it had dropped to under 1 million—an unfathomable loss of more than 95 percent of the population.

Shrinking indigenous populations led colonizers to take actions that amplified the impact of epidemics. In Peru, Viceroy Francisco de Toledo instituted a resettlement program in the 1570s, taking families from formerly sparse settlements and forcibly concentrating them into denser villages (called *reducciones*). This order, intended to facilitate colonial oversight of indigenous bodies, also dramatically increased their population density, situating them in a "death trap" when measles and smallpox epidemics arrived several years later. Thus colonizers not only brought disease-causing microbes to the Americas but also created conditions that enhanced contagion and its consequent fatality among susceptible native populations.

In many cases, colonizers did not have to take any action at all to gain advantages from the ongoing outbreaks. The viruses did their work for them. Epidemic disease "ran ahead of direct contact" with Europeans, destabilizing populations in the future southeastern United States, the Mississippi River valley, and elsewhere—setting the stage for easy plunder. Puritan settlers, previously exposed to measles and smallpox in the Old World, arrived to find empty fields and deserted settlements and were only too happy to claim this contagion-cleared land as their own.

Colonists were grateful for the perceived blessing of massive depopulation of indigenous people by infectious disease. A 1643 publication titled *New England First Fruits* conveys that message clearly, numbering smallpox first among a long list of divine gifts. In their minds, God cleared a path for their progress, and the chosen implement was contagion. One of the most prominent American colonists, the Puritan minister Cotton Mather, shared a similar view. Mather was born in Boston, the son of the minister of the Old North Church, and spent a lifetime preaching and engaging in public discourse. Among his many pamphlets and book-length works was the *Magnalia Christi*

Americana, a mammoth seven-book ecclesiastical history of New England. In the first book of the *Magnalia*, Mather explained the role of pestilence in removing the "pernicious creatures" who were in the Puritans' way.

> The Indians [*sic*] in these parts had newly, even about a year or two before, been visited with such a prodigious pestilence, as carried away not a tenth but nine parts often (yea, 'tis said, nineteen of twenty) among them: so that the woods were almost cleared of those pernicious creatures, to make room for a better growth.

Although Mather, ostensibly a man of God, saw contagious diseases as a means to divinely ordained success for the colonists, his family was not spared from the ravages of such illnesses. In 1702, the year that the *Magnalia* was published, an outbreak of smallpox spread through Boston, sickening three of his children. His wife, Abigail, who had lost a child due to premature birth only months before, died during the outbreak.

That outbreak was not an anomaly. Columbus and the settlers who followed spent their childhoods in areas endemic for Old World diseases, maximizing the difference in susceptibility between indigenous and colonizing groups. However, as the colonies thrived, more children were born on colonized land, far from European cities, increasing the susceptible local population with every birth. Ships continued to cross the Atlantic. Thus, as the American colonies grew, both measles and smallpox caused intermittent and often lethal outbreaks among the non-indigenous population.

In 1713, epidemic disease visited Cotton Mather's home again. In mid-October, Mather's diary entries demonstrate his clear apprehension about the "common Calamity of the spreading measles." On October 18, his son, Increase Mather, fell ill. Within weeks, the entire

Mather household—his second wife Elizabeth, their newborn twins, the seven older children, and a "maidservant"—was ill with measles. Helpless, Mather recorded in excruciating detail the illness and recovery of some of the children, the passing of the maidservant due to a "malignant fever," and his anguish at the deaths of his wife, the twins Eleazar and Martha, and two-year-old Jerusha.

That particular wave of measles lasted far beyond the dates of the tragedy in the Mather household. Boston held official citywide days of fasting and prayer over the next several months to ask for deliverance from the scourge. Measles continued to spread within Boston and to other parts of the colonies. Determined to take some action in the face of what must have seemed a hopeless situation, Mather composed a letter urging Bostonians to take measles seriously. His first line reads: "The Measles are a Distemper which in Europe ordinarily proves a Light Malady: but in these parts of America it proves a very heavy Calamity; A Malady Grievous to most, Mortal to many, & leaving pernicious Relicks behind it in All." This disease is, for obvious reasons, no joke to him, and perhaps by pointing out the difference between its behavior in Europe and in America, he hoped to convince newer arrivals of its severity. On the other hand, it could have been just a strong lede—early eighteenth-century clickbait. The substance of the text is interesting to a modern reader, in part for the peculiar juxtaposition of the raging outbreak—enhanced by Mather's firsthand experience with the measles's toll—with the quaint-sounding remedies that he suggests (sage tea, syrup of saffron, maybe a little mulled cider).

Several demographic trends shifted both the frequency and severity of measles in the North American colonies throughout the 1700s. Metropolitan populations grew, and commerce and travel intensified. That meant that measles outbreaks might be more extensive when they did occur, but because of the efficient spread of the virus, fewer susceptible individuals would remain once cases finally waned. The second trend involved international commerce. Over time, trade and travel between the colonies and European ports increased. More ships meant more potential importations of measles, particularly if the ships carried children, the most likely sources of the virus. Improved ships meant more efficient transatlantic travel. Measles would still have to go through a chain of two susceptible hosts to make it across the ocean, but maybe not three. Together, increased domestic population size and regular importation of measles from Europe by ship meant more frequent outbreaks in a progressively less susceptible population.

This shift, from small to large population density and from sporadic explosive outbreaks to more frequent, milder ones was the beginning of measles's change from an epidemic to an endemic disease in the colonies. By the end of the century, larger metropolitan areas in North America had at least small numbers of measles cases in most years, punctuated by occasional larger outbreaks. Of course, not everywhere was Boston or Philadelphia, and when measles did arrive in more remote colonial towns after long absences, it could be just as devastating as ever.

Not all places in the world experienced this shift of measles to an endemic disease at the same time. In the 1800s, as British colonialism expanded and ship technology improved, new areas—particularly the island nations of the Pacific—saw their societies upended as measles came to their shores. The disparity between the toll that measles takes on large, connected populations compared to smaller, isolated ones is seen most clearly in island outbreaks. The same pattern that Peter Panum detected in the Faroes—an extensive outbreak sparked by travel from a continental city to an island with a small population and marked geographic and economic isolation—played out on multiple Pacific islands in the 1800s, with vast human and political consequences.

Having explored the South Pacific extensively over the prior century, the European powers, particularly the British, formally annexed many Pacific Island nations in the 1800s. Colonial expansion provided local strategic advantage, access to abundant natural resources, and new, exploitable pools of labor. As in the Americas, measles facilitated the Europeans' success in obtaining these spoils.

In October 1874, King Cakobau of Fiji formalized a deed of cession, transforming the islands into a British colony. The British representative during the negotiation and signing of that document was Sir Hercules Robinson, the colonial governor of New South Wales, Australia. In December 1874, Cakobau, who maintained a position of authority in the new colony, along with an entourage that included two of his sons (Ratu Timoci and Ratu Josefa), traveled to Sydney for a state visit with Robinson. Unfortunately, that visit coincided with the early stages of a massive measles outbreak in Eastern Australia. With increased ship traffic in the preceding decades, once-rare importations of measles became more frequent in Australia in the 1800s, and the 1874–1875 outbreak was particularly deadly. Most reports focus on the considerable toll measles took on the European colonizers, but death rates from measles were even higher among the aboriginal peoples. The presence of measles in Sydney in 1874 would also prove fateful for Fiji, as the former king became infected during his state visit. Cakobau recovered, although press coverage of his return to Fiji weeks later commented that he still looked weak on arrival. The ship's physician diagnosed both Ratu Timoci and another Fijian traveler with measles en route. Based on the length of the voyage, a little over three weeks, Cakobau himself could not have been contagious by the time the British HMS *Dido* returned to the Fijian capital of Levuka on January 12, but some passengers had visible rashes upon arrival.

Prior to cession, Fiji had no recorded experience with measles. No rules were in place for quarantine or inspection of incoming ships to prevent importation of infectious diseases, and the prevailing custom was for private boats to meet and immediately board incoming vessels to greet travelers and exchange news. Shortly after the annexation of Fiji, Dr. John Cruickshank, a retired British naval officer, became the provisional chief medical officer, but a lack of facilities and support staff meant that he had not activated British colonial quarantine procedures by the time the *Dido* returned. Compounding the problem, neither the colony's new governor nor the permanent chief medical officer had even arrived in Fiji by the time that the outbreak began.

The maritime sign for a ship carrying ill passengers with smallpox, measles, or other potentially infectious diseases is a yellow flag. Hoisting that sign upon arrival to the harbor would have signaled that welcoming vessels should not approach the *Dido*—that isolation would be required for ill passengers and crew and quarantine for those who were exposed. No such flag flew at Levuka.

Cakobau and his sons returned to the family home at Draiba, just outside of Levuka. Within days of their arrival, Ratu Josefa and his attendant developed the measles, indicating exposure on board the ship. Despite this development, Cakobau convalesced and prepared for a multiday event at which more than five hundred people from throughout Fiji came together to feast, share news, and discuss life under colonial rule.

The subsequent months have been called Fiji's "darkest hour." Measles exploded, spread throughout the islands by festival attendees. Cases and deaths were widespread by mid-February. The colonial government imposed quarantine regulations in late February, too late to stem the tide of the epidemic. The toll of measles was sudden, ubiquitous, and disastrous throughout Fiji. The disease affected individuals of all ages, and so many lay ill or dying that societal structures were fractured. Vacant supply ships bobbed in harbors. The fatality rate among indigenous Fijians was so high that burying the dead became impossible. First mass graves and then sporadic shallow burials were the best that many areas could accomplish. Farmers could not harvest crops, and parents were unable to care for their children. Starvation weakened even those who had survived the initial attack of measles, and secondary illnesses like dysentery and pneumonia followed. Measles depopulated whole families, in some cases entire villages, within weeks.

Weather and societal disarray worked in synergy with measles. Storms brought more than fifty inches of rainfall over the course of a month. Downstream effects on sanitation and the availability of safe food and water increased the toll of the epidemic. Local government was upended. Though Cakobau survived, he lost a son, a daughter, and a brother. Regions lost their political leaders, along with a large percentage of the adult population. The overall effect has been called a "decapitation of society."

Fiji's new status as a British colony both initiated and exacerbated the destruction. Fijian adherence to public health recommendations regarding quarantine, isolation, and care of people with measles was limited at best—a downstream effect of distrust of the newly arrived authorities. Theories that Europeans had purposefully brought measles into the islands and that non-native medications might be poisonous flourished. The sick abandoned British-run hospitals, refusing offers of medicine and food. Many descriptions of the situation come from British reports, in which a tone of colonial haughtiness is apparent. However, it is easy to empathize with feelings of fear and distrust among the Fijians, as fresh off the signing of the deed of cession, measles arrived from another British colony with disastrous results. *The Fiji Times* espoused exactly that view: "The first advantage derived from annexation is the introduction of measles, and for this we are indebted to the *Dido*, which discharged her diseased passengers utterly regardless of any consequences that might arise." The political situation directly affected Fijians' experience of the measles epidemic and understandably limited the trust that they were willing to place in the British public health authorities.

Local records are incomplete, but the epidemic waned by early May, and new cases ceased by mid-June. Supplies returned, and recovery began. It is difficult to determine mortality rates accurately, but some areas reported that nearly half of their population had died during the outbreak. The official estimate for measles-related deaths during the 1874–1875 epidemic in Fiji is about 40,000—more than a quarter of the total population lost. Even once the acute outbreak had ended, late complications of measles and prolonged effects on the demographics of the native Fijian population persisted. Direct mortality and indirect effects, including decreased birth rates, ensured that Fiji's population would remain destabilized for many years.

While Fiji's outbreak was remarkable in both pace and scope, its fate was not unique among the islands of the Pacific in the 1800s. Measles executed a more focused decapitation of Hawaii's government, killing both king and queen during an 1824 visit to England and foreshadowing a tragic outbreak that would devastate its population two decades later. Late in 1848, overlapping outbreaks of measles and pertussis swept the islands, leading to tens of thousands of deaths (estimates range from 10 to 30 percent of the total population) and broadscale disruption of Hawaiian society. The initial introduction of measles came from the *Independence*, an American naval vessel traveling from Mexico.

The havoc that measles brought to Fiji, Hawaii, and other islands in the 1800s made clear that this disease posed an existential threat to formerly isolated nations that were then joining the "global pathogen pool" through colonialism and commerce. Such trends accelerated as European and American colonization expanded and the efficiency of travel grew. As in Mesoamerica, measles caused deaths, but it didn't act alone. Colonialism, political aggression, and distrust all increased the virus's destructive power, leaving weakened indigenous populations less able to resist their intruders.

5.

Crowded, Poor, Malnourished

In April 2023, the commissioner of New York City's Department of Health and Mental Hygiene sent a letter to local health care providers. Over the prior year, more than fifty thousand people seeking asylum had crossed the U.S.–Mexico border and were relocated to New York City. The health needs of this group were vast, complex, and evolving. In addition to reminding providers to screen for trauma, developmental disabilities, evidence of human trafficking, lead exposure, and a host of other factors, the commissioner emphasized the specific risks due to infectious diseases in this community. Outbreaks of chickenpox had sickened many children and young adults, tuberculosis was on the rise, and a substantial percentage of children had not received vaccines against COVID-19 and other preventable diseases, putting them at ongoing risk.

It was helpful for the commissioner to bring more attention to the challenges faced by migrant families. However, for many of us, caring for children who had journeyed from South or Central America, often traveling with their families by foot for incredible distances under inhumane conditions and then housed here in packed shelters, had become part of our daily professional lives. Infectious diseases thrive under these conditions—when people are crowded, frightened, and malnourished. I was part of a team who took care of an adolescent with untreated tuberculosis—a CAT scan showed the top of one of his lungs hollowed out into a cavity—as he struggled to breathe in a hospital room. All five of his family members crowded around the bed as his mother explained her fear that their temporary shelter, once a medium-priced midtown hotel, would force the whole family out if they learned of her son's diagnosis. For months over the course of their journey, he had symptoms—fever, cough, night sweats, weight loss—but his mother hadn't been able to help him. There was no way to see a doctor, have an X-ray, or get medications. The deprivation and the physical and mental strain of the voyage exacerbated what was already a serious infection. By the time he was finally able to get medical care, his condition had become far worse—and the required treatment far more extensive—than it ever should have been. What could have been cured with a prescription for antibiotics months earlier now required scans, surgery, and a long hospitalization to get under control.

In addition to untreated tuberculosis, we have seen recently arrived children with dengue fever, malaria, undiagnosed and untreated cancers, and a host of other infectious and noninfectious health problems. We have been fortunate not to see cases of measles in this migrant population in New York City so far, but if we do, there is the potential for disaster. The crowded quarters, tenuous nutrition, chronic stress, and lack of access to vaccines provide a perfect situation for measles to spread, as an early 2024 outbreak in Chicago has demonstrated.

Measles not only loves a crowd, it needs a crowd. It never could have successfully spilled over into humans without one. The story of how measles found us and how it held on to become a quintessential human pathogen lives at the intersection of viral biology and human behavior.

The measles virus is a member of a family called the paramyxoviruses, which have RNA-based genomes. Like measles, many of the other paramyxoviruses cause disease—mumps and parainfluenza are two examples. The paramyxovirus subgroup that contains measles virus is called the morbilliviruses, and while measles affects only humans, other morbilliviruses have their own preferred hosts. These include canine distemper (dogs), rinderpest (even-toed ungulates such as cattle), and the beautifully named peste des petits ruminants (goats, sheep, other ruminants, and camels). Using genomic sequencing techniques, scientists have recently clarified the evolutionary relationships among the morbilliviruses. As a result, we now have an idea of when and how measles virus—or a close ancestor—may have made its jump into humans.

Phylogenetic trees are pictorial representations of how organisms are related, much like a family tree for humans. For viruses, these trees are based on DNA or RNA sequences and are useful tools to understand viral evolution. The morbillivirus tree has canine distemper virus branching off to form its own separate subgroup. The remaining branch represents the common ancestor of peste des petits ruminants virus, rinderpest virus, and measles virus. These three are closely related, but peste des petits ruminants separates first, leaving measles and rinderpest to share a common ancestor. Thus, while measles is a uniquely human disease, unmatched in contagion and spread, its closest cousin isn't a virus that causes a similar human disease like smallpox. It's rinderpest—the cattle plague virus.

When combined with additional information about the organisms themselves (factors like preferred host species, the time and place that the sample was collected), phylogenetic trees allow us to reconstruct a plausible version of history—to estimate dates and sometimes places for important events, including species divergence, emergence of pathogens, and host switches, including spillover events. Using this approach, for example, DNA sequences from *Helicobacter pylori*, a bacterial species that can cause ulcers but can also persist for long periods of time in the intestines without causing symptoms, revealed ancient human migration patterns, including those driven by colonialism, slavery, and war. Studies of ancient genomes of *Yersinia pestis* (the cause of bubonic plague) from Bronze Age archaeological sites in Russia led to the realization that this flea-transmitted plague may have arisen more than a thousand years earlier than previously thought. In modern outbreaks, similar techniques have enhanced our understanding of the emergence and spread of Ebola in West Africa and the spillover of SARS-CoV-2 from bats to humans.

Phylogenetic approaches have also been used to understand the evolutionary history of the morbilliviruses, including the separation of measles and rinderpest viruses from their common ancestor, but this work is challenging. RNA viruses generally have high mutation rates compared to DNA viruses and nonviral organisms. Extensive mutation can lead to less predictable evolutionary trajectories and can make estimating divergence dates difficult. In addition, unlike the Bronze Age plague genomes or the DNA virus genomes that implicated the transatlantic slave trade in global viral spread, ancient RNA virus samples are extremely hard to come by. RNA is unstable in the environment—much more so than DNA. Fluctuating temperatures, changes in humidity, and just plain time all lead to physical decay of RNA. A vast viral history may go undetected because scientists can analyze only modern samples.

There have been sporadic successes in recovering RNA viruses from decades or centuries ago. The scientific rewards of such work can be great—even paradigm-shifting—but most attempts end in failure. In the 1990s, scientists managed to isolate fragments of the RNA genome of the 1918 pandemic influenza virus from long-preserved laboratory samples and from a corpse recovered from Alaskan permafrost. Those samples provided a wealth of information about how viruses evolve and how pandemics begin. Did the 1918 virus kill tens of millions of people because it was particularly good at damaging human cells? Or because it transmitted from one person to another more efficiently than other influenza strains? Or because it was different enough from previous flu viruses that huge numbers of people became ill all at once? These previously unanswerable questions were suddenly tractable, all because of scientists' newfound ability to pry genetic information from long-dormant tissues.

That technical triumph wasn't simply the result of someone having the idea to try to resurrect an old virus. It required two separate lines of research to converge with just the right timing, a story that began nearly half a century earlier. In 1950, Johan Hultin was a medical student from Sweden who had come to the University of Iowa to do graduate work in microbiology. During lunch with a visiting professor, the topic of the 1918 influenza pandemic came up. The professor suggested that to understand that catastrophe, someone ought to go "find bodies in the permafrost that are well preserved and that just might contain the influenza virus." Hultin, inspired by the remark, proposed the idea to his thesis advisor and began to assemble information. He contacted Otto Geist, a paleontologist at the University of Alaska, and wrote to several mission sites that had suffered many deaths during the 1918 pandemic. By mid-1951, Hultin, his thesis advisor, and a pathologist went to Alaska, met up with Geist, and journeyed to Brevig Mission (called Teller Mission at the time). Brevig Mission had lost approximately 90 percent of its population during the pandemic, and local officials had to bring in gold miners to bury the bodies in mass graves two meters deep in the permafrost. The village council gave their permission for the expedition to dig at the site, and after several days' work Hultin discovered the body of a young girl buried in a gray dress. The next day's work revealed four more bodies. Clad in surgical masks and gloves, the team removed chunks of lung tissue from each body. They used carbon dioxide from fire extinguishers as a substitute for dry ice to preserve the samples as they left the site, beginning their return trip to Iowa. Back in the laboratory, Hultin tried every technique he knew to try to revive the influenza virus from those precious samples, but none was successful. Their effort had been heroic, but it had ultimately failed. This may very well have been a good thing, as containment procedures at that time were not sophisticated, and there was a chance that the Iowa scientist's mission could have unleashed the pandemic influenza once again.

The story of the resurrection of the 1918 flu might have ended there—an exciting idea felled by the constraints of the real world—but Hultin never gave up on the idea that there were important scientific truths hidden in the permafrost. In 1995, Dr. Jeffery Taubenberger used newly developed techniques to extract fragments of RNA from preserved autopsy tissues in the archives of the Armed Forces Institute of Pathology (AFIP). His team identified samples from thirteen cases from the 1918 pandemic that they thought had a good chance of yielding enough RNA for analysis. One of the thirteen turned out to be positive. Although the genetic information that Taubenberger recovered was highly fragmented, it was enough to allow him to publish a manuscript that began to address crucial questions about the pandemic virus—for example, that it was closely related to modern influenza strains. This paper represented an important step forward, but the small RNA fragments coaxed from the preserved autopsy samples were not enough to put together the full sequence of the virus, which would be key to understanding where the 1918 influenza came from and what made it so deadly.

Hultin, still interested in the secrets of the 1918 virus, read Taubenberger's paper and wrote to him. He described the 1951 Brevig Mission work and offered to return to the site to gather additional tissue. Taubenberger was interested, and within a week, Hultin began a self-financed return to Alaska. The mayor and village council allowed him to dig at the site again. This time, he found a corpse of a young adult who had died from influenza and whose lungs were well preserved. Hultin collected tissue samples—no fire extinguishers this time, though he did use a set of garden shears borrowed from his wife—and sent them to Taubenberger, who found abundant viral RNA in Hultin's samples, an important step up from the minuscule amounts in the AFIP autopsy specimens. Hultin's newly collected tissue allowed the long and eventually successful march toward a complete decoding of the 1918 virus to move forward. As a result, we now understand much more about the 1918 virus than we ever could have without Hultin's work, including the fact that when it emerged it was likely a brand-new virus for humans, originating from a strain that circulated in waterfowl. That jump from one species to another was a turning point in human history, not just because of the 1918 pandemic that followed, but because every influenza pandemic and even most seasonal influenza infections since then have come from descendants of that 1918 virus.

More recently, similar techniques led to reconstruction of a complete human immunodeficiency virus (also an RNA virus) genome from a 1966 sample collected in the area that is now the Democratic Republic of the Congo, helping to clarify the evolutionary origin of that pandemic. Even a single sample with well-preserved RNA can make scientists rethink when and how a viral family evolved. As the story of Johan Hultin shows us, recovering a decades-old or century-old RNA virus is an incredible accomplishment requiring a combination of tenacity, luck, and skill. As the result of another recent success in this area, scientists have unlocked similar insight into the origin of measles.

Sometime around June 1, 1912, a two-year-old girl was admitted to Charité Hospital in Berlin suffering from measles complicated by pneumonia. After three days in the hospital, she died. Even in its highly sanitized and technical form, her autopsy report is wrenching to read. Childhood deaths are difficult to process, even across the centuries, and particularly so when the deaths are due to conditions that we can now prevent or cure. The report describes a "well-muscled" and "well nourished" child with a red throat, swollen tonsils and lymph nodes, highly abnormal lungs with evidence of pneumonia, and a rash. One of the child's lungs was preserved in a specimen collection that had been started by Rudolf Virchow (Panum's friend—see chapter 3) and maintained by his successors. It remained in storage there for more than a century.

In late 2018, Sébastien Calvignac-Spencer, a virologist at the Robert Koch Institute in Berlin, identified the girl's preserved lung as a sample that, with careful analysis, might yield genetic sequences that could help clarify the timing of the emergence of the measles virus. His team took the specimen to their laboratory, a carefully monitored area designed to avoid contamination with DNA and RNA from the modern world. They removed small pieces of tissue from the lung, stained some for microscopic analysis, and isolated RNA from the others. The images of this hundred-year-old tissue under the microscope appear much as a measles-infected lung would look today—areas of pneumonia where inflammation and white blood cells crowd out the normally open airways, giant cells with multiple nuclei. The real information bonanza came from the RNA sequencing data. Despite decades in formalin preservative, the team found that the RNA quality was better than expected, and sequencing yielded a nearly complete measles genome.

Combining that 1912 viral sequence with more modern samples, Calvignac-Spencer's group constructed a phylogenetic tree, yielding an estimate for the date of the measles-rinderpest divergence—528 BCE, more than twenty-five hundred years ago. The date may not be accurate to the year, or even the century—the uncertainty range that the scientists calculated around their 528 BCE estimate is still about six hundred years in either direction—but the timing makes sense in the context of information we have from other sources, including Rhazes's writings about measles and smallpox from the ninth century. The date also fits our understanding of early human societies and the population size that measles requires to keep a chain of infection going.

The critical community size for a specific pathogen to achieve endemicity depends on a number of factors. Transmission efficiency is one factor, as are the length of time that an infected person is contagious and whether infection reliably leads to long-lasting immunity. For measles, that population threshold is somewhere in the range of 250,000 to 500,000 people. Based on what we know about the history of human societies, that puts an outer limit on how old measles can be. Early human settlements had nowhere near that number of people, and even if migration or trade allowed smaller areas to connect (and share infections) with one another, crossing the quarter-million threshold was a relatively recent event.

The rise of agriculture and domestication of animals approximately ten thousand years ago allowed human populations to expand to unprecedented densities. Those innovations also fundamentally changed the relationship between humans and some animal species, including cattle, putting them in close, sustained contact and enhancing the potential for pathogen spillover. Modern estimates have multiple communities crossing the threshold during the first millennium BCE, a time consistent with the Calvignac-Spencer estimate for measles emergence. Exceeding that population threshold fundamentally changed the interactions between morbilliviruses and people, allowing measles to remain in human populations, to begin its long arc of constant coexistence with our species.

This is a point with modern resonance as well, as changes in the way that humans interact with the environment, such as urbanization, can increase spillover events with potentially fatal viruses. One striking example is a close relative of the measles virus called the Nipah virus. Nipah is also a member of the paramyxovirus family. It is transmitted from animals—mainly bats or pigs—to humans, and can occasionally be transmitted from person to person by close contact within families or in hospital settings. Nipah virus outbreaks are highly lethal and have become larger and more common in recent years. The burden of Nipah is particularly evident in areas like the Southern Indian state of Kerala, where ongoing deforestation has disrupted bat habitats and increased human-bat contact. Kerala has had at least four Nipah outbreaks since 2018, with multiple deaths.

As more and more cities exceeded the measles critical community size, they experienced changes in both who got the measles and how the public perceived the disease. London had become measles endemic by the late 1700s, but most surrounding areas had not, due to their smaller populations. The American colonies had at first behaved like islands, but by 1830, the mortality bills of New York City listed hundreds of measles deaths every year, evidence of endemicity. Some years had spikes in case numbers, but even during the off years, measles was never absent. The other major eastern port cities of the United States—Boston, Philadelphia, Baltimore—had all surpassed the measles critical community size by the 1870s, and some cities of the interior such as Chicago and St. Louis followed shortly thereafter. The actual burden of measles in these metropolises in the late 1800s and early 1900s is impossible to quantify, as most cases did not require medical attention and centralized reporting was generally not mandatory. In contrast to the rare but explosive outbreaks in isolated islands, for urban-dwelling families measles became a ubiquitous part of childhood.

Consistent with that trend, in 1913, Dr. William Butler made an attempt to scientifically evaluate the "popular assertion that everyone must have measles." Working in the Willesden District northwest of London, Butler surveyed families of children who had been recently "exposed intimately" to measles, asking about the children's prior history. He also recorded which of these kids went on to develop measles after exposure. What he found was remarkable. Of the nearly 14,000 children for whom he collected data, nearly 80 percent had already had the measles. When broken down by age, the data showed that while less than 10 percent of the youngest group (those in the age range of zero to four years) had a measles history, more than 97 percent of those over age fifteen did. Butler's additional findings validated the family members' recall. Most children with no measles history went on to develop the disease after their exposure, whereas the ones who had had measles before did not. Butler's work suggested that the popular conception of the ubiquity of measles was essentially correct, at least in Willesden. By the time they reached young adulthood, nearly all children had had measles and were thus protected from getting the disease again. Endemicity pushed the age distribution of measles cases toward the young, and the result was an adult population with lifelong protection.

That shift to a younger age distribution applied to measles fatalities as well. While outbreaks in isolated populations led to deaths across the age spectrum, after that initial exposure, even if the population remained too small for true endemicity, subsequent measles fatalities were mainly concentrated in young children, those between six months and two years of age. Butler noted, as others had before him, that children under the age of six months rarely got the measles, and when they did, their cases tended to be mild. This observation was, seemingly paradoxically, also the result of the transition of measles into a childhood disease. Because nearly all mothers had had measles in childhood, they could pass antibodies to the fetus during gestation, providing strong, though temporary, protection for the earliest months of life. Today we take advantage of this kind of protection by vaccinating pregnant people against influenza, pertussis, COVID-19, RSV, and other diseases, with the goal of providing early life protection against these important infections.

Because the vast majority of children recovered from measles without incident and because urban parents understood that their children would get it sooner or later, there was also a substantial shift in how the disease was perceived. London still had considerable measles mortality rates (around 5 per 1,000 children under age five annually at the turn of the century), yet its very ubiquity reinforced the idea that coming down with the measles was "a necessary part of the game." Doctors lamented that parents did not take the disease seriously. An 1896 article titled "Murder by Measles" cites an old saying that "every man must catch measles and fall in love." A cartoon in the popular London-based satire magazine *Punch* shows a well-dressed mother holding a tray with a serving dish on it, about to enter a child's room as her other child looks on. The caption reads, "'Mummy, may I have the measles when Violet's finished with them?'" The implication is that a bout of childhood measles is a time for rest and pampering, rather than an experience with a potentially fatal disease. Something to be envied, not feared.

While it was essentially true that there was no avoiding measles as a child in London or New York City, its perception as a mild disease by the middle and upper classes of society was also driven by marked socioeconomic disparities in measles mortality. It wasn't simply that wealthier families were blind to childhood deaths due to measles. It was that their children were considerably less likely to die of the disease than those of poor families.

Such disparities were discussed only sporadically. In 1885, Harry Drinkwater, a general practitioner, tended to cases during a particularly virulent measles outbreak in the city of Sunderland in northeast England. Between mid-January and late March of that year, he cared for more than three hundred people with measles and tabulated his detailed observations as a handwritten thesis. Drinkwater found the outbreak unusual in the speed of its onset and spread, but his most striking conclusions involved the distribution of cases, especially severe ones. He listed sections of the town where measles had been most prevalent and where "the great majority of deaths have occurred" and concluded that the underlying cause "does not lie in any specially defective sanitary arrangements, but in the great poverty of those residing therein." In contrast, he reported that cases in the "well-to-do" part of town were of the usual type, with few deaths as a result. He attributed the high rates of severe complications and deaths in the impoverished areas to malnutrition, which he called "semi-starvation," previous respiratory infections, and homes with multiple cases during the outbreak, explicitly connecting social and economic conditions with disease outcome. Drinkwater submitted his thesis to the University of Edinburgh for completion of his doctoral degree and garnered the Gold Medal prize for his work.

Drs. Waldo and Walsh, the authors of "Murder by Measles," pointed out the stark differences in measles death rates between two similarly named London districts in 1894, one wealthy (St. George's, Hanover Square) and the other poor (St. George's, Southwark). The latter district had triple the rate of measles deaths of the former. Combining these findings with overall death rates from other infectious diseases in London, they concluded that "In measles we find a highly infectious disease, which, although comparatively harmless among the rich, plays havoc among the poorer classes of society. It is looked upon by most folk as a malady that calls for little or no treatment." Waldo and Walsh also took the public health system to task for failing to mandate reporting of cases of measles. In their view, compulsory reporting of cases would allow a better understanding of the disease's behavior in the population as well as possibilities for early action, particularly keeping exposed and infected children home from school during the period of contagiousness.

Income and measles severity were linked, but exactly how poverty might translate into worse disease was not obvious. Drinkwater's work suggested a role for malnutrition. Later studies would confirm that relationship and also pinpoint vitamin A deficiency as a particularly potent risk factor for severe measles. In addition, one of the most important determinants of measles outcomes is the age at which a child is infected. The youngest children with measles—predominantly those in the critical period between six months and two years of age—are considerably more likely to develop complications such as pneumonia or to die from the infection. Therefore, factors that cause children to acquire the infection during that vulnerable window also increase the overall death rate from measles. It's not just whether a child gets the disease, it's when.

In the 1920s, James Halliday, an officer of the Public Health Department in Glasgow, turned careful observation into a theory that linked poverty with measles mortality. Butler's interviews had in part laid the groundwork for Halliday's ideas. In Butler's data, everyone got the measles at some point, but the children of the "more favored classes" were less likely to be infected during early childhood than their poorer counterparts. This observation fit with a growing understanding of the importance of socioeconomic disparities in health as well as urban schools as potential sites of spread of epidemic diseases. Halliday extended these lines of thought. He realized that early life exposure to measles among poor children could be both a cause of increased mortality (as these earlier cases were more often fatal) and a direct result of their housing conditions. He tested this hypothesis through a detailed study of Glasgow tenements. Halliday noted, "The outstanding characteristic of a tenement building is that a number of families are congregated around a close [i.e., an entryway] or stairway through which all the inhabitants must pass. In the poorer districts, the landings, close, and stairways are also used as a play-ground by the younger children." The result of the tenement architecture was congestion, loss of privacy, and an inability to effectively isolate even ill children.

Even in the tenements, Halliday found that parents considered measles an annoyance rather than a true threat. Children with mild cases of measles, though still contagious, roamed the hallways. Halliday mapped the location of first (index) cases in specific tenements as well as each of their contacts, categorized by floor and apartment position. This strategy, though cumbersome, revealed that many measles cases started at school, where children from different neighborhoods mixed, and subsequently spread through the tenements—not only to the index case's siblings but to their neighbors. Through these routes, susceptible preschool-age children living in tenements became secondary cases nearly four times as often as their older counterparts.

The same was not true among better-off families. Comparing tenement infection rates to those from "housing scheme" areas (those with less crowded dwelling structures) and to rates outside of Glasgow led to a striking conclusion—among families living in tenements, measles became a disease of early childhood rather than school age due to exposure patterns mandated by the structure of the buildings. These early infections in turn conveyed an increased risk of complications and death, simply by virtue of the age at which they occurred. Poverty, by way of architecture, dictated health.

The Glasgow tenements were not unique. Journalist and social reformer Jacob Riis published his masterwork, *How the Other Half Lives*, in 1890. Riis documented the horrors of poverty in New York City tenements through text and photographs. Decades before Halliday's investigation, Riis intuited the difference in what measles meant for the rich and the poor and linked this dichotomy inexorably to tenement life.

> There were houses in which as many as eight little children had died in five months. The records showed that respiratory diseases, the common heritage of the grippe and the measles, had caused death in most cases, discovering the trouble to be, next to the inability to check the contagion in those crowds, in the poverty of the parents and the wretched home conditions that made proper care of the sick impossible.

Newer data indicate that crowded housing may increase the risk of death from measles for reasons beyond just exposure at an earlier age. This evidence suggests that intensity of viral exposure, driven by crowding, poor ventilation, and a host of other factors, can also be a determining factor in who lives and dies from measles. The narrow hallways and landings of tenements and the lack of well-ventilated rooms intensify exposures, making it more likely that a young child will not only get measles earlier than she might otherwise have, but that her risk of that infection progressing to bronchitis, pneumonia, or even death will be amplified as a result of the increased number of virions that she inhales.

Crowding also makes measles a lethal disease among adults, often illuminating social inequities. In the American Civil War, "camp measles" and the pneumonia that frequently followed it led to illness and death in large numbers of Union and Confederate soldiers. Crowded conditions and poor nutrition in barracks and in prison camps accelerated viral spread and infection severity among all troops, but those factors were substantially worse among Black soldiers. An analysis of more than 75,000 cases of measles among Union troops revealed more than 5,000 fatal cases, an enormous toll. The fatality rate among white soldiers with measles was about 6 percent; for their Black counterparts, it was nearly 11 percent.

Measles preys on the weak, the crowded, the malnourished. The children in the Glasgow tenements died because their families' poverty led them to live where they would get exposed to measles earlier and with greater intensity. Blacks died at nearly twice the rate of whites in Civil War camps and in colonial Philadelphia not because of intrinsic susceptibility but because of malnutrition and inhumane living conditions.

Here is Riis again, understanding that it is not absence of money that makes measles lethal—it is that money buys the things that tenements lack: space, air, and light.

> Here is a door. Listen! That short hacking cough, that tiny, helpless wail—what do they mean? They mean that the soiled bow of white you saw on the door downstairs will have another story to tell—Oh! a sadly familiar story—before the day is at an

end. The child is dying with measles. With half a chance it might have lived; but it had none. That dark bedroom killed it.

Pediatricians like me focus on individual children. We see the patient in front of us, taking note of the fluid behind their eardrum that might indicate an ear infection, the peculiar character of their gait that might tip us off to a possible cause of their abdominal pain. We do our best to diagnose accurately and treat effectively and safely—to make that particular child better. That kind of interaction is a crucial piece of pediatrics, but it is not the full picture. My colleagues and I also think about, and in some cases spend our careers studying, social determinants of health—all of the factors that feed into who gets sick and who stays healthy, who thrives and who doesn't. For an individual, there can be elements of randomness—a child with an allergy takes a bite of the wrong piece of food; another is on a street corner at a tragically wrong time. Learning to consider populations as well as individuals broadens our perspective. When we carefully count not only who gets sick, but where and when, we learn that more factors play into health than we might have originally thought—factors like class and geography and the distance from a child's home to the nearest supermarket. The history of measles teaches us that political decisions and neighborhoods and nutrition and home ventilation all matter.

We have known for hundreds of years about the intersection of money and colonialism and racism and living conditions with the toll of infections. This tangle of causality isn't gone or even going—it's just getting started. This way of thinking is not just relevant to contagious diseases. Heart disease and diabetes and car accidents and shootings are about much more than the involved individuals. The history of measles is an allegory, as there is nothing like an epidemic to bring disparities into sharp relief.

Not long after the first COVID-19 cases appeared in the United States, the differences in its impact on rich and poor America and on white and Black America became clear. It wasn't just case rates that were higher in poorer, Blacker areas—death rates were higher, too. At the peak of the pandemic, we saw this clearly among the kids we admitted to the hospital with COVID-19. These were frequently the children of low-paid but essential workers—restaurant employees, delivery drivers, transit workers, and others—parents who could not stay home and isolate, some of whom felt pressure to work even if they felt ill. Factors that could buffer some of the impacts of COVID-19 on families—an extra room in which to isolate an ill family member, ventilation to decrease the risk or intensity of exposures, the option to work from home, a functional internet connection so that a child could attend school remotely—were, and are, inequitably distributed. As measles should have warned us, so were, and are, the worst outcomes of the COVID-19 pandemic.

6.

Making Nothing Happen

Prevention can be a tough business. A pediatrician talks to a parent about choking hazards or sleep position or bike helmets, but she never gets to know which specific children that advice has helped. You can't see prevention unless you broaden your view, looking at populations over time. Getting rid of leaded gasoline decreases childhood lead poisoning; changing recommendations about infant sleep positions lowers the risk of sudden infant death syndrome. But you don't get to know which kids benefited—who would have not worn a helmet and had that bike accident, tipped over that unsecured television stand, died of SIDS. Prevention is invisible. It's a very different kind of medicine from taking care of a child in an ICU with the flu or even a kid in your office with an asthma attack, where the focus is on the most specific details of that child in front of you. Pediatricians are lucky enough to be able to do both. Vaccines are one of our best tools for prevention. They are amazing inventions that prevent serious diseases so kids can get on with their lives. But you never really know exactly who they helped. Vaccines are masters of making nothing happen.

We get the measles only once. Despite the clever ways that the virus eludes our defenses, co-opting the cells of our immune system, measles infection leads to lifelong immunity. Our immune system processes, remembers, and protects. For that reason, a child with measles poses no threat to an adult who had it when they were young. This aspect of measles biology is understood by physicians and grandparents alike.

This state of affairs is mostly good news, as adults who contract measles are at higher risk of complications than all but the youngest children. At a population level, that means that the overall toll of measles is much lower when the virus continuously circulates than when outbreaks are infrequent but explosive. The catch is that we pay a price for the endemicity that brings that decrease in severity. A population with near-universal measles infection in childhood must tolerate a constant burden of rare but severe cases, including some deaths. The paradox is that even as measles becomes a childhood rite of passage, every year that it circulates, some children will die of the disease. Others will develop life-altering complications—deafness, blindness, neurological diseases. Measles infection leads to brain inflammation (encephalitis) in about one out of a thousand cases. One out of every ten thousand children with measles develops subacute sclerosing panencephalitis, a progressive and fatal neurological disease that begins several years after the original infection. A continuous hum of rare tragedy is the price of endemicity.

For a long time, there was no choice but to accept this situation, as no vaccine was available to prevent measles until the 1960s. This chapter tells the story of that vaccine, which represented a fundamental shift in the way humans and measles interacted and, for the first time, provided hope that a third choice could be added to the endemic/epidemic dichotomy—elimination.

It was not for lack of trying that measles continued to take its yearly toll in the pre-vaccine era. The idea that doctors could provide specific protection against some diseases—making them optional instead of mandatory—was not a new one. Protection through inoculation, the purposeful introduction of material from someone with a disease into the body of a susceptible individual, had been around for centuries. It was first practiced as a means to prevent smallpox infection. That procedure, called variolation, took several forms, all of which had a shared goal. If all went well, variolation induced a mild case of smallpox in the recipient—some skin lesions and fever, but none of the life-threatening aftereffects. Lifelong protection without danger—the holy grail. All did not always go well, though, and because the material used in variolation contained potentially virulent smallpox virus, the procedure could cause severe and sometimes fatal smallpox infections.

The earliest written reports of variolation originate from China, where practitioners of the technique would grind smallpox scabs into powder and blow them into the nose of an uninfected person, a procedure called insufflation. In India, practitioners obtained material from skin lesions using a needle and inoculated it into the skin of another person—a strategy that more closely resembled the method that would eventually come to Europe and the American colonies in the early 1700s. West African nations developed similar approaches, likely over the same time period.

In 1714, Emanuel Timonius wrote to the Royal Society from Constantinople to report on a new (at least to him) method for smallpox prevention. Timonius described the procedure and said that "thousands of subjects for these eight years past" had undergone successful variolation. Despite the importance of smallpox as a cause of disease and death, Timonius's report did not trigger immediate widespread use of variolation in England, but it did prompt a written response from a celebrity on the other side of the Atlantic.

Cotton Mather, the Boston preacher and recent survivor of the measles epidemic that had devastated his family, was generally supportive of Timonius's contribution, though with a bit of a backhanded edge. After bestowing his "favorable Opinion" on the letter, Mather claimed priority for knowledge of this unusual medical procedure. Mather was a slave owner who, horrifyingly, had received a West African man, whom he named Onesimus, as a "gift" from his congregation in 1707. In his response to the Royal Society, Mather reported that he had asked Onesimus whether he had ever had smallpox. Onesimus answered "both yes and no." Pressed for details, Onesimus educated his enslaver. Mather wrote that Onesimus "had undergone an Operation, which had given him something of the *Small-Pox*, & would forever preserve him from it . . . He described the Operation to me, and shew'd me in his Arm the scar, which it had left upon him." And thus, on the basis of stolen knowledge from a stolen life, steeled by his family's own tragedy and perhaps urged on by Timonius's report, Cotton Mather positioned himself to become America's champion of variolation. Mather advocated the technique during a major smallpox outbreak that started in Boston in 1721, overcoming skepticism and overt hostility—most notably when a lighted grenade crashed through his window. It did not explode, but its accompanying note read, "Cotton Mather, You Dog, Dam you: I'll inoculate you with this, with a Pox to you"—an early example of the lengths to which anti-vaccine individuals would go to attempt to silence and intimidate others.

During the same period, England had its own highly visible proponent of the variolation technique, Lady Mary Wortley Montagu. Lady Mary had traveled to Constantinople with her husband, Edward Wortley Montagu, who was appointed ambassador to the Ottoman empire in 1716. She had a long-standing and quite personal interest in smallpox—she had lost a brother to the disease and was badly scarred after her own case. Shortly after their arrival, Montagu learned of variolation ("The small-pox, so fatal, and so general amongst us, is here entirely harmless by the invention of *ingrafting*," she wrote to a friend) and had it performed on her young son, Edward. After her return, she promoted the procedure in England, eventually convincing Caroline, Princess of Wales, to allow her own children to be inoculated. Despite the royal imprimatur, the procedure remained controversial in London. Over the next several years, favorable reports from the American colonies and systematic collection of data from other countries helped tip public opinion toward variolation.

Given the heat of these debates, it is curious that several decades passed before reports appeared describing application of the inoculation technique to infections other than smallpox. In 1754, *The Gentleman's Magazine* printed a letter from T.S. describing a method for preventing "distemper" among horned cattle, a condition caused by the rinderpest virus, measles's closest relative. T.S. attributed the procedure to a Mr. Dobson. Dobson's proposed intervention owed a substantial debt to variolation. In essence, it involved making an incision in the loose skin around the neck (the dewlap) of a healthy recipient animal and inserting fibrous material that had been dipped in secretions from the nose or eyes of an ill animal. "[L]et it remain there two or three days, in which time the distemper will appear; then turn the beasts out into dry pasture, and let them remain there until the crisis of distemper is past." In so doing, Dobson reportedly saved "nine in ten of his whole stock." Following this creative expansion of inoculation as a preventive strategy, several other writers suggested that measles, and potentially other diseases such as plague, might be next.

In 1758, Francis Home, an Edinburgh physician, took the bait. A measles epidemic had begun in Edinburgh late in 1757 and continued well into the next year. Although he described the disease as generally mild, Home attended numerous fatal measles cases. For some of these, he collected notes, including recordings of patients' vital signs and detailed descriptions of their breathing. He published these in a book titled *Medical Facts and Experiments* as a prelude to describing his own attempts at measles inoculation. He justified his decision to try this untested procedure in humans in this way,

> Considering how destructive this disease is . . . considering how many die . . . considering how it hurts the lungs and eyes; I thought I should do no small service to mankind, if I could render this disease more mild and safe, in the same way as the Turks have taught us to mitigate the small-pox.

Home's hypothesis was that if he could find a way to transfer the infection "by the skin alone," he might be able to avoid the more severe downstream effects of measles, particularly pneumonia and other complications that involved the lungs. He could not get enough material from the arms of a measles patient by rubbing them with a woolen glove, so he decided to turn to "the magazine of all epidemic diseases, the blood." Home drew blood from measles sufferers at the height of their illness, making an incision in a superficial vein in the area where the measles rash was thickest, soaking the blood into cotton, and then, as soon as possible, beginning the inoculation process. For the recipient, he made an incision in each arm, allowed it to bleed for fifteen minutes, and then put the cotton inside, leaving it there for three days.

Home inoculated twelve children with this procedure, and then for an additional three, he tried a different tactic—inserting cotton with either nasal secretions or blood from a measles patient into the nose of a child. For each, he kept track of their symptoms over the following days. Nasal transmission was ineffective, but he found that most of the children who received inoculation via arm incisions developed a mild form of measles about a week after the procedure.

The most remarkable thing about Home's experiments with measles inoculation isn't so much that he made the attempt but his near silence on the matter after that initial report. Other than describing one child from his original group, who went on to develop the measles again about six weeks after inoculation, he did not report whether the children were protected from future attacks. After the initial descriptions in his book, he mentioned the procedure only briefly in another of his works and reported no further trials. It is easy to be critical of Home's work in retrospect, to point out that measles was circulating in the community and that the mild cases that some of his recipients developed may have been just that—mild cases of regular measles acquired in the normal way. Yet these trials still represent the first attempt at purposeful, specific protection against measles, the beginning of the long road to a vaccine.

Over the following decades, others attempted inoculation against measles with variable success. Public reaction to the idea of measles prevention using this technique ranged from enthusiastic to dismissive. It remains unclear whether these skeptical views resulted from distrust of Home's findings, a lack of enthusiasm for employing this procedure to prevent a disease that was often mild, or a combination of the two factors. Notably, both Peter Panum and August Manicus knew about Home's work prior to their investigations in the Faroe Islands. They elected not to inoculate any of the Faroese against measles, as they presumed that nearly everyone had already been exposed by the time they might have been able to do so. Panum also expressed concern that inoculation might make matters worse by promoting spread of the infection.

In general, the subsequent reports of measles inoculation, whether or not they were successful, were either direct attempts to replicate Home's approach or minor variations on that theme. That changed on April 28, 1915, when Charles Herrman, a pediatrician on Manhattan's Upper West Side, addressed the annual meeting of the Medical Society of the State of New York. Herrman began by acknowledging that the mortality from measles had been declining in New York in recent years, but reminded his audience that it was still a serious disease that caused complications frequently. "On several occasions, I have stated that this high morbidity could only be distinctly diminished by immunization against this disease. For several years, I have had a method in mind, but it was only about two years ago that I gathered courage to test the method."

Herrman had read about the recent discovery of a "filterable agent" (an early term for a virus) in the blood and nasal secretions of humans with measles that could cause a measles-like illness in monkeys. He also knew the prevailing theory that children under six months of age were protected from measles, provided that their mother had had the disease prior to pregnancy, but that the protection would wear off over time, leaving them vulnerable. Only that wasn't always the case in his experience. He presented an instructive case to his audience. A nearly four-month-old child was exposed to an older sibling with measles but did not become ill. As the family grew over the years to a total of seven children, measles swept through in several waves, but that particular child never developed measles. Herrman theorized that the early exposure, toward the end of the protected period, had engendered long-lasting immunity without any sign of illness from measles. He hoped to harness the idea of intentional transmission of measles but to do so in that protective time window. In his words, the goal was to "convert a temporary into a more or less permanent immunity"—the long-term protection of having measles without the risk.

So Herrman collected snot. His ideal donor was a child in the prodromal phase of measles, the hours leading up to the eruption of the rash. He harvested mucus from the nose and stored it on cotton swabs in glass jars. Once he had parental permission, he would apply one of the swabs "gently to the nasal mucus membrane" of a healthy four- or five-month-old infant. At the time of the 1915 conference, he had performed forty of these inoculations, with no ill effects. Four of the forty infants had been closely exposed to measles in their families more than six months after the inoculation, and none had gotten the disease. He "challenged" an additional two with a second swab exposure as they approached their second birthday, and neither contracted measles.

Herrman continued his work, and when he returned to update the society in 1922, he had performed the procedure on 165 New York City children and had monitored nearly half of them for more than four years. Of those, only 5 had gone on to develop measles. In addition, he had performed his rechallenge experiment on 36 children, none of whom had become ill. In his view, this outcome represented an unalloyed success.

This new approach to measles prevention seems to have worked. Aside from the ick factor when viewed retrospectively, when he performed the initial inoculations, Herrman was working within the ethical bounds and scientific norms of the time, and he did a reasonable job of systematically collecting data and reporting his findings. The challenge aspect of the experiments (i.e., exposing a group of previously inoculated children to potentially infectious secretions for a second time) is more dubious by modern standards. In his mind, this second exposure was likely just a logical extension of the experiments. Inoculate, then challenge—how else to determine whether the procedure provided protection? The problem is that second exposure, given after protection from maternal antibodies would have waned, was not done with the intention to protect. These were children who, if the initial inoculation had not worked (or had worked only imperfectly), could have suffered measles and its attendant risks as a direct result of the challenge procedure.

Despite the successes of Herrman's research, there was limited excitement for nasal inoculation of infants, either in the United States or elsewhere. Reliance on a supply of snot from children with measles at just the right stage of illness did not make for an easily scalable solution, and some parents might have been understandably reluctant to have another kid's boogers put into their child's nose.

In Herrman's time, intentional exposure of children to measles was already going on, just in a more haphazard way. One of Herrman's stated rationales for his nasal inoculation procedure was to provide an alternative to the age-old custom "to intentionally expose children to infection with measles so that it would be over and done with." He saw that homegrown strategy as significantly riskier, particularly for children young enough to be at higher risk of complications but old enough to have lost their early life protection.

Other authorities spoke out against these "measles parties" as well. Just before Thanksgiving in 1913 and next to an article about a scam perpetrated by a phony bandmaster, the *Bridgeport Evening Farmer* ran a three-column piece by the assistant surgeon general of the United States, Dr. William Colby Rucker. The article's overwrought title, "Measles Makes Many Mothers Mourn," was followed by a plea to parents to take measles seriously and not to purposely expose their children. Rucker tells the story of Little Johnnie, who catches the measles by playing "post-office" with girls at a party one Saturday afternoon. Little Johnnie is miserable with the measles, but luckily his case turns out to be mild. Unfortunately, Little Mary, who lives next door, is sent over to Johnnie's house by her mother, who hopes that Mary will get the measles so that she can be done with them by "bridge season." Mary, predictably, does not fare well in this scenario, and her death from measles is presented as a cautionary tale for mothers (fathers are not mentioned in the article).

Despite the sermonizing tone of Rucker's writing, he was making a reasonable point. Measles was still a major cause of disease in the United States at that time, and it was minimized by parents, some of whom did send their children to measles parties (or over to Little Johnnie's house). And Rucker believed what he was saying, repeating similar messages, though without the homilies, in his writings for medical audiences as well. His 1916 government report on the measles concluded: "Measles kills more people in the United States every year than smallpox. You can't kill a child any deader with smallpox than you can with measles. It is the duty of private citizens and municipalities to take every known measure for the prevention of the spread of this disease."

Unfortunately, given the combination of measles's contagiousness and its ubiquity, other than discouraging parents from purposefully exposing their children to people with the infection, there were not many effective preventive measures for them to take. Reliable protection of children from measles would require vaccination, and despite the attempts of Home, Herrman, and others, there were still substantial technical hurdles to achieving that goal.

In June 1939, John Enders, a forty-two-year-old assistant professor at Harvard, spoke at a symposium dedicated to reviewing current knowledge of some important viral diseases. His topic was "The Etiology of Measles." Enders was, in many respects, an improbable guest at the gathering. He had been born to a wealthy family in West Hartford, Connecticut. His grandfather was a president of the Aetna Life Insurance Company; his father, the head of the Hartford National Bank. Enders attended elite schools (St. Paul's School, Yale College) and served in the Naval Reserves. After college, he had tried a career in real estate but found it unfulfilling. Enders's initial experience with graduate school in Celtic and Teutonic languages at Harvard was little better. He bounced from one thesis topic to another without finding passion.

During this time, Enders shared a house with several other students and young faculty members, one of whom was an instructor in the Department of Bacteriology and Immunology at the medical school, studying under the eminent professor Hans Zinsser. Enders accompanied his friend to the laboratory on occasion and became excited about the work. He switched his PhD concentration from languages to biology and joined Zinsser's group. Even within this new realm, Enders remained something of a dilettante. Tuberculosis, pneumonia, cat viruses—all were fair game in his early years in the lab.

At the symposium, Enders reviewed what science knew about measles, starting with Home's report and continuing through the list of human and animal studies that followed. He concluded that measles was probably caused by a virus, though conclusive proof was still lacking, and he bemoaned the absence of a solid, reproducible experimental animal model of the disease. He had tried to induce measles in rabbits, guinea pigs, a "fairly large series" of mice, rats, two ferrets, chicken eggs, and monkeys. His egg experiments yielded some intriguing but inconclusive results. The monkey studies were plagued with inconsistency.

Some trials seemed to be successful, with the monkeys developing fever, rash, and other signs of measles, and others yielded no reaction at all.

Consistent with his prior behavior, Enders moved on to other research topics, but the measles mystery remained with him. Over the next few decades, his work on another disease, polio, would make him wildly successful—a scientific hero, a Nobel Prize winner—and he would then use the products of those other labors to revisit and finally to resolve the measles conundrum.

Polio vaccine development was a legendary triumph of the 1950s, and the names that most people associate with that accomplishment are scientists—Jonas Salk, Albert Sabin. These two took fundamentally different approaches to the same problem. Salk's inactivated polio vaccine and Sabin's live vaccine both provided protection against an epidemic microbial scourge that stoked fear throughout the world. There were certainly no widespread "polio parties" where parents intentionally exposed their children. Salk and Sabin famously and continuously feuded over whose approach to vaccination was better. In truth, we are still fighting the polio eradication battle, and both tools have been indispensable to the incredible progress that has been made.

The other truth is that neither Salk's nor Sabin's vaccine would have been possible without the work of John Enders. In 1946, Enders moved his laboratory from Harvard to Boston Children's Hospital. There, Enders and two junior researchers, Thomas Weller and Frederick Robbins, developed new methods for propagating viruses in tubes coated with thin layers of human or animal cells. Viruses can grow only within cells, and a standardized method for generating large amounts of virus is a key step in developing a vaccine. To generate the tubes of cells, they collected organs, minced them, and separated individual cells from one another using specialized enzymes. The single cells were collected in new dishes and fed with a liquid nutrient medium. Weller was particularly interested in varicella virus (the cause of chickenpox), and

Robbins focused on viral etiologies of diarrhea. Both used the cell culture technique to grow viruses, combining patient samples with the cell layers and watching them under the microscope over the course of days or weeks, changing the liquid every few days and looking for cytopathic effects—the telltale signs of cell damage resulting from viral replication.

Occasionally, there were leftover vials. Enders, famously frugal, would not allow the precious cell layers to go to waste. Even though the prevailing dogma was that poliovirus would grow only in nervous system cells, which were extremely challenging to culture, he suggested testing some polio samples in the extra vials. The results of these early offhand experiments changed everything—the focus of Enders's laboratory, the career trajectories of these three men, the pace of polio research across the globe, and the way that vaccines were (and would be) made. Poliovirus grew in cells derived from discarded foreskins from circumcisions. It grew in cells from embryos. It grew in cells from intestines.

Getting poliovirus to thrive in cell layers threw open a door. The team produced large amounts of virus, from which vaccines might be developed. Comparing the extent of infection in cells exposed to poliovirus in the presence or absence of human serum allowed detection of poliovirus-specific antibodies. Careful dilutions of those serum samples revealed the point at which the protective effect faded—the basis for a quantitative assay to monitor the rise and fall of an individual's antibody levels over time. Enders shared both samples and knowledge from his laboratory freely, accelerating the field toward effective polio prevention. Salk and Sabin each received both viral samples and technical advice from Enders's group—essential early steps on the paths to their vaccines. Accolades followed. Weller and Robbins left to become department chairs. Enders was elected to the National Academy of Sciences, and the three shared a Nobel Prize in 1954 "for their discovery of the ability of poliomyelitis viruses to grow in cultures of various types of tissue." This was the only Nobel awarded for the polio vaccine, and it went not to Salk or to Sabin—the warring vaccine developers—nor to Dr. Thomas Francis Jr., who carried out the massive initial field trials of Salk's vaccine. It went to Enders and his lab mates, whose tubes, cells, and persistence had made it possible.

A colleague would later recall that, since the beginning of his work on viruses, Enders had found measles more interesting than polio. Polio work continued in the Enders laboratory after the Nobel Prize, but that success also allowed him to return his attention to the unfinished business of measles. It was time to try out some of the cell culture techniques that had worked for polio on his old nemesis. Enders offered this project to a new research fellow in the group, Dr. Thomas Peebles. Peebles had a peripatetic career reminiscent of Enders's own. As an undergraduate, he majored in French and played the snare drum in a band. He served as a Navy bomber pilot in World War II and had considered accepting an offer to fly for a commercial airline once the war was over. He turned that opportunity down and pursued medical school instead, followed by training in pediatrics at Massachusetts General Hospital. As a pediatrician, Peebles had seen firsthand the importance of measles as a disease of children. He accepted Enders's offer.

Peebles's first step was to try to grow measles virus in several different types of cell culture. The team had the cells, but getting fresh samples of blood and mucus from children with measles would be key to their success. After some false starts, a colleague, Dr. Thomas Ingalls, provided a suggestion that ultimately proved fruitful. Ingalls, an expert on rubella, had worked with Enders a few years earlier to investigate an epidemic of respiratory infections at a boys' school. They had combined epidemiologic investigations with antibody testing performed in the Enders laboratory to understand the dynamics of the outbreak and, crucially, to pinpoint its cause (influenza A). This time, Ingalls heard about a measles outbreak at a different local boarding school. He alerted

Peebles, who saw the potential to not only identify cases of measles as they happened, but to get samples from the children back to the lab quickly, a crucial factor in successfully coaxing viruses to grow.

Peebles spoke to the school's principal about the project and obtained permission to identify boys in the early stages of the disease and to invite them to take part in the research. He appealed to the boys' sense of pride at participating in a scientific investigation: "Young man, you are standing on the frontiers of science." Participation in the study meant collection of a blood sample, a throat "washing" (obtained by gargling with a tablespoon or so of sterilized skim milk and spitting it back into a container), and a sample of feces. Peebles would then rush each of these samples back to the Enders laboratory and inoculate them into vials as soon as possible.

Peebles's first success came from case number three, an eleven-year-old boy identified in the report as D.E., David Edmonston. On February 8, 1954, David had a low-grade fever and some abdominal cramps. He had been in contact with several other boys who had developed measles. By the next evening, a faint facial rash and a rising temperature made the diagnosis of measles almost certain, so Peebles was summoned. David later recalled, "He asked if I would help the world, so I said yes, of course."

As had been the case with polio, there was no obvious recipe for growing measles virus. Peebles proceeded by trial and error, using a panoply of cell types as potential hosts: the first lineup included human kidney, embryonic lung, uterus, even cells grown from monkey testicles. On February 10, he inoculated David's samples into the vials. The kidney cells, which were included in the group only because Enders happened to know a surgeon who could provide a steady supply, were the clear winner. On day five—just seven days since David Edmonston's fever and cramps—multinuclear giant cells, clear evidence of viral cell damage, emerged in one of the tubes.

True to his nature, Enders was skeptical at first. The cellular changes did not resemble the cytopathic effect that his team had seen with polio and other viruses, but Peebles knew the significance of the giant cells as a sign of measles infection and held his ground. They reproduced the effect by taking a small amount of fluid from the infected kidney cells and inoculating a fresh batch of cells with it ("passaging" the virus), time after time after time. Staining the cells with specialized dyes revealed a "lace-like network" of infection. Finally, when they adapted the polio antibody test approach, mixing in serum from people who had recently had measles, their antibodies inactivated the virus and prevented giant cell formation. At that point, even Enders had to admit that Peebles probably had done what he set out to do. He had tamed measles virus, getting it to survive and reproduce in captivity.

There was still the problem of the monkeys, though. Measles virus grew efficiently in cultured kidney cells from either humans or nonhuman primates. However, Enders's preliminary experiments suggested that some monkeys did not get sick when exposed to measles, but others were susceptible. The team's newfound ability to grow measles in cell culture helped solve that mystery as well. It turned out that the monkeys who did not get infected had antibodies against measles in their serum (likely from prior exposure to humans with measles), while those whose antibody testing was negative were fully susceptible. The laboratory-passaged virus caused a measles-like illness in those seronegative animals, complete with rash in some cases. Crucially, those symptoms preceded a robust antibody response, protecting the monkeys against a second round of infections. The laboratory-cultured virus gave the animals measles—but only once. Just like humans.

This is the point at which Enders's approach to measles diverged from the one that he had taken to polio. He had learned his lesson from the endless and exhausting battles between Salk and Sabin and took on the challenge of developing a measles vaccine himself. The Edmonston virus that grew in kidney cells seemed perfectly capable of causing disease in primates, as long as they didn't have antibodies already, but a vaccine would require separating the pathogenicity of the virus (its ability to cause illness) from its immunogenicity (its capacity to engender a protective immune response). One approach, à la Salk, would be to grow large amounts of virus and inactivate it with formaldehyde or a similar chemical. The other, more Sabin-like and the one that Enders favored, would be to create an attenuated version of the measles virus—one that could still grow in cell culture and lead to production of antibodies when given to people but that was weakened enough not to cause disease.

Additional members of the research group tried to grow the virus in different cell types. Enders suggested that the group attempt to coax the virus into more distantly related cells, with the idea that it might lose some pathogenic properties along the way. That *might* would mean the difference between a safe vaccine, one that could induce its recipients to make a protective immune response without exposing them to danger, and no vaccine at all.

Twenty-four passages in human kidney cells gave rise to a viral strain that—unlike the original sample—could divide in human amnion cells (the cells that make up the membranes that surround a developing fetus). Twenty-eight passages in those cells unlocked robust growth in chicken embryos. Six in the embryos led to success in the simpler system of cultured chicken cells. Fourteen in chicken cell culture, and the team felt ready to test whether this system of repeated passaging had produced a virus that was less pathogenic. Along the way, the virus had clearly changed. Its range of possible host cells had greatly expanded, and there was a new cytopathic effect—long, thin "spindle cells." This was the Edmonston A strain. For good measure, the team made a B strain as well, putting it through several additional rounds of passage in both chicken embryos and chicken cell culture. It still wasn't clear that any of these procedures meant that an attenuated virus would be the end result, or even if they had achieved that goal, that the weakened virus would remain measly enough to lead to durable immunity.

Both things ended up being true. While it remains unclear exactly how or when in its long, multi-species journey the Edmonston strain lost its ability to cause the symptoms and complications of measles, that is exactly what happened. Vaccination of monkeys with either strain did not lead to fever, rash, or any of the other signs of measles, but after a few weeks, the vaccinated animals had developed high levels of anti-measles antibodies, as measured with the serum assays. The important final step was to then challenge those vaccinated monkeys with the real thing. So, more than five months after vaccination, they received the non-attenuated version of the Edmonston virus.

The greatest test of a vaccine is its ability to make nothing at all happen. That is the blessing and the curse of vaccination. When it works well—when we get everything right—there is nothing to see. For a healthy child, the absence of illness isn't usually dramatic—it just looks like regular life. It's only in a controlled setting like a laboratory experiment, when you know the timing and the dose of an exposure, that not getting sick is a thrilling and measurable event. The results of the challenge experiments were exciting in exactly that way. The vaccinated animals didn't get sick; the vaccine had protected them. Joyful news, but it meant that there was now a lot more work to do.

After the successful vaccine trial in monkeys, there was only one logical next step. Enders's new vaccine had to be tried in people. Self-experimentation has a long history in infectious diseases research and particularly in vaccine development. Maurice Hilleman, the legendary Merck scientist who brought dozens of vaccines to market over the course of his career, has summed up this approach as "you make it, you take it." The Enders group was no exception to this philosophy. They produced standardized, sterility-tested, and storage-stable lots of experimental vaccine, and then the team members, all of whom had had measles in the past, injected one another with the attenuated vaccine. As they had hoped, nothing much happened.

Measles vaccination was not meant for already-immune adults, though. Eventually, a first measles-naive child would have to receive the vaccine. This is another point at which the ethical norms of the 1950s, 1960s, and 1970s are at clear odds with modern approaches to research with human subjects, particularly children. In his comprehensive biography of Hilleman, *Vaccinated*, Dr. Paul Offit addresses the strange dichotomy that existed in vaccine research at this time. Vaccine developers such as Hilleman frequently tested their candidate products on three groups of people—themselves (and often their colleagues), their own children, and disabled children residing in institutions. Often, though not always, the conditions in such institutions were horrific, and there were frequent outbreaks of measles, mumps, infectious hepatitis, and other conditions that the investigators were interested in studying. Often, though not always, the investigators obtained consent from the parents or guardians of the children as well as the leaders of the institution.

Contemporaneous as well as later interviews with scientists that led such work tend to emphasize their interest in protecting children from the recurrent outbreaks of diseases that occurred in such institutions. Measles and other infections spread rapidly among the young, crowded, and frequently undernourished residents, and complications like pneumonia were common. And yet the fact remains that these were experimental vaccines, neither thoroughly tested nor licensed for use. Consent was often cursory, and parents may not have been informed fully of the goals of the research, the specifics of the procedures, or the attendant risks.

Notably, the other groups involved in early vaccine testing at the time (scientists, their colleagues, and their children) were also susceptible to coercion. It may have been understandably difficult for junior researchers or the children of a scientist to question or refuse an offered vaccine. In the late 1970s, Hilleman's employees at Merck were informed by the boss himself that there was no option for refusal of the hepatitis B vaccine that he was developing.

Dr. Samuel Katz, a pediatrician, had already received the attenuated Edmonston virus with the other members of Enders's group, when he decided to vaccinate his own children with it. However, to assess the safety and efficacy of the vaccine, he needed access to a lot more kids. Enders suggested that he contact the leaders of the Walter E. Fernald State School in Waltham, Massachusetts. The Fernald School—originally founded as the Massachusetts School for the Feeble-Minded in the 1840s—had grown to embrace the philosophy of the eugenics movement. Fernald himself, the school's third headmaster, was a member of the board of directors of the Eugenics Society, advocating separation from the general public and forced sterilization of those whom he considered genetically inferior. Those attitudes continued among subsequent administrations, and the institution has a notorious and well-documented history of physical abuse of its residents and participation in unethical experimentation on children. Among the worst of these were a series of studies in the 1940s that exposed children to oatmeal laced with radioactive isotopes without adequate parental consent.

In comparison, the experiments that Katz and Enders proposed to the Fernald administration and the parents of the involved children were relatively tame. Measles was a problem at Fernald, with significant outbreaks every few years, and there was reason to think that the vaccine would confer protection. Given the conditions of the institution and the ongoing neglect, Fernald residents were at high risk of complications of measles. The investigators had taken reasonable precautions in terms of confirming sterility of the vaccine doses and testing them on both experimental animals and some adults prior to use. Parents gave written consent. Even so, the fact that these were neither the first nor the worst of the studies performed on institutionalized children in that era does not make them easier to consider with a twenty-first-century eye.

The fortunate conclusion to the measles vaccine work at Fernald was that it was both safely completed and a clear success. Thirteen children, ranging in age from two to ten years, participated in the study. None had had measles before. The Enders lab tested their serum, ensuring that each child lacked anti-measles antibodies. On October 15, 1958, Katz injected eleven Fernald children with Enders's vaccine. Two others received a placebo as a control.

By the end of the study, two things were apparent. First, most of the children who received the vaccine rather than the control preparation developed fever starting about a week after vaccination, sometimes followed by a rash that looked like a mild case of measles. Second, all the vaccinated children (but neither of the controls) developed high levels of anti-measles antibodies within a few weeks of the injection, evidence that they might be protected from infection.

Katz presented these results at a meeting of the Society for Pediatric Research in May 1959. His talk received a place of honor in the first plenary session of the meeting, and lively discussion followed. There was clear interest in the prospect of measles vaccine development, but the issue of vaccine acceptability by both parents and physicians was already on the minds of the audience members. Dr. C. Henry Kempe had already begun an early trial in collaboration with Katz and Enders, using the same vaccine, and he anticipated issues with vaccine demand that might follow its eventual licensing. He commented that mothers rarely thought of measles as a threat. In addition, there was concern that the side effects of the vaccine might make parental acceptance less likely. Several days of fever was hardly inconsequential for a young child or their parents, and discussions began in earnest about the feasibility of further attenuating the vaccine virus while retaining its effectiveness. In a public lecture several months prior to the Katz presentation, Enders himself had reported some of the early findings from the Fernald study and suggested that more attenuation might be required in order to make measles vaccination practical.

This early work drove intense interest and a flurry of clinical trials of the Enders vaccine over the subsequent few years. It was clear to Enders and the others in the group that they would need help from other researchers to bring the attenuated measles vaccine through the large-scale testing that would be needed to gain approval. For that reason, the Enders group shared the attenuated virus with many other scientists, some from academic laboratories and many from pharmaceutical companies. Merck's Hilleman was one of those who collected the attenuated Edmonston virus, setting about his own measles vaccine research with his signature focus and intensity. Enders also reached out to his friends and colleagues for help. Robbins from his polio days, C. Henry Kempe, Francis Black from Yale, a Boston Children's Hospital pediatrician who ran a large community pediatrics program, and Saul Krugman from Bellevue Hospital in New York City. Each set to work testing the vaccine in different populations of children.

The July 28, 1960, issue of *The New England Journal of Medicine* was an unusual one—so much so that the editorial board felt their readers deserved an explanation. While the journal generally covered a range of medical topics each week, the development of a new measles vaccine was important enough to merit eight separate papers, taking up nearly the entire issue. The issue included an introductory article by Katz, Enders, Milovanovic, and Ann Holloway, a technician in the Enders group. That description of the early studies of the Edmonston strain, its attenuation, and the preparation of vaccine was followed by six articles, each detailing one of the clinical studies carried out by Enders's network of friends, beginning with Katz and Holloway's Fernald School study. Finally, a summary article made the case that, based on the results from the 303 children enrolled across all six studies, the attenuated measles vaccine appeared to be both safe and immunogenic. They recommended larger scale trials but also noted that parents would have to be prepared for the "expected reactions to the vaccine," which included fever in most children and rash in nearly half.

Work on the vaccine continued on many fronts. There were still significant issues to address prior to its approval. Hilleman, convinced that Merck should be the company to produce the measles vaccine, remained concerned about the frequency and the intensity of the side effects. In an interview in later years, he called it "toxic as hell." He worked on the issue with Joseph Stokes Jr., a Philadelphia pediatrician who had a theory. Gamma globulin, a preparation of pooled antibodies from the blood of adult donors, generally had high levels of anti-measles antibodies. This was the result of measles being a near-universal childhood disease. It was already in use for measles, though in a different way. Children who had a known exposure to someone with measles could be given an injection of gamma globulin, which would often lead to a milder "modified" case of measles. Stokes and Hilleman showed that giving low doses of gamma globulin along with the Enders vaccine greatly decreased fever and rash without disrupting the immune response to the vaccine.

The gamma globulin strategy addressed one of the major drawbacks of the candidate vaccine, but it also made administration unwieldy. Stokes's approach required two shots, and the dose of gamma globulin was important. Too much, and the effectiveness of the vaccine was compromised. Too little, and the fever and rash broke through. In addition, there was significant variability in the amount of anti-measles activity in gamma globulin from different production lots, complicating attempts to standardize dosing. The gamma globulin approach represented an improvement over the Enders vaccine alone, but it was hardly ideal. For that reason, some researchers were losing faith that a live-attenuated measles vaccine could be successful. Several companies, including Merck itself, were testing killed-virus vaccines based on the Edmonston strain, raising the possibility of another Salk-vs.-Sabin-like "war" between proponents of the two strategies.

In early November of 1961, the National Institutes of Health (NIH) convened a conference to organize and discuss the state of this rapidly growing—and increasingly chaotic—field. There was clear governmental interest in licensing one or more measles vaccines. The International Conference on Measles Immunization brought together academic, industry, and government representatives to review the available data and begin to carve out a path to that goal. President John F. Kennedy, who was occupied by a visit from the prime minister of India, sent a letter to the conference chairman expressing his regrets at not being able to address the group in person and lauding the "leading scientific minds of 20 nations," who had gathered "to protect children against a formidable and widespread threat."

The opening remarks from Surgeon General Luther Terry emphasized the global burden of measles disease and set his expectation for the assembled experts, that their work over the subsequent few days would provide a road map leading to the eventual approval and use of a measles vaccine. The conference was remarkable not only for its "everyone who's anyone in measles research" cast but also for the breadth of the presentations. Reports on the importance of measles came not just from Alexander Langmuir, chief of the Epidemiology Branch of the Communicable Diseases Center (the precursor to today's Centers for Disease Control and Prevention, or CDC), who focused on the United States, but from teams from India, Nigeria, Brazil, Chile, the USSR, Greenland, Tahiti, Australia (with a notable focus on the disease's potential impact on indigenous people), and South Africa as well. Langmuir, for his part, put forward a remarkable proposition: that with the tools that were becoming available, control and "early eradication" of measles could soon be at hand.

Studies of live-attenuated vaccine came from Enders's and Hilleman's groups, of course, but also from Israel, England, Japan, and other countries. The live vaccine had been tested in children with cystic fibrosis, heart disease, asthma, tuberculosis, and leukemia. Killed-virus vaccine studies came from groups at Pfizer and Eli Lilly. Their vaccines required multiple doses, and the antibody responses waned over a few months, but gamma globulin was not needed, and fever and rash were substantially rarer.

There were discussions of animal models, viral culture techniques, and manufacturing practices. Roderick Murray from the Division of Biological Standards of the NIH led off a session anticipating potential problems in measles vaccine production and standardization. He expressed concern about the need to evaluate three distinct products—live-attenuated vaccine, killed vaccine, and a standardized preparation of gamma globulin. One—maybe more than one—measles vaccine was headed rapidly toward U.S. approval, and regulators wanted the kinks worked out as early as possible.

On March 21, 1963, Secretary of Health, Education, and Welfare (HEW) Anthony Celebrezze Sr. formally announced the U.S. licensure of two measles vaccines, Pfizer's inactivated vaccine (Pfizer-Vax Measles-K) and Merck's live-attenuated vaccine (Rubeovax), the latter to be given along with Gammagee, Merck's standardized gamma globulin preparation. Surgeon General Terry held a press conference to trumpet the accomplishment. Terry emphasized the severity of measles in the United States—four million cases and hundreds of deaths annually—with an even steeper toll abroad and suggested that the vaccines could help bring about measles elimination within two years.

Notably, the surgeon general's announcement did not come as a surprise to the press. There had been considerable coordination among the U.S. Public Health Service, members of the press, and representatives from the involved companies for months leading up to the announcement. In addition to a mid-February 1963 announcement to writers and editors that the approval would happen soon, PHS staff worked with representatives from the Columbia Broadcasting Service for well over a year on a planned television special about measles. *The Taming of a Virus* was hosted by Charles Collingwood and featured Drs. Enders, Peebles, and Hilleman. *The New York Times* critic called it a "quietly compelling drama" and was particularly moved by the segment that described the emergency use of live-attenuated measles vaccine in West Africa, in a campaign in which more than 700,000 children were vaccinated in an attempt to stem a deadly outbreak.

One aspect of the approval announcement received only minor attention, though it had an important, perhaps decisive, effect on the success of early measles vaccine rollout. Surgeon General Terry advised that children aged nine months or older who had not yet had measles should get vaccinated, but that no large-scale public vaccination drives were planned. The government preferred that private physician offices and clinics distribute measles vaccine. Despite the Kennedy administration's interest in promoting child health, particularly through federal support of vaccination, it seemed that there would be no immediate plan for central coordination or financing of this particular vaccine.

To add to the confusion, although two vaccines were approved, the CDC's advisory committee expressed a clear preference for the live vaccine, except for people with medical conditions that precluded its use. There was logic behind this particular recommendation—antibody responses to killed vaccine were weaker and shorter-lived than those to the live-attenuated strain—but Terry's announcement described the substantially lower rates of side effects for the killed vaccine, a feature of potential interest to parents. Both enthusiasm for and availability of the killed vaccine were limited given the tepid language of the recommendation, but its inclusion as a potential option confused both parents and physicians.

For these reasons, the public response to measles vaccine's unveiling represented a substantial contrast to the polio vaccine's triumphant arrival almost eight years earlier. The results of the large-scale trial of the Salk vaccine in children—those 1.8 million children who volunteered to receive either vaccine or placebo were called "polio pioneers"—had been nationwide, even global, news. The positive study results were greeted with jubilation, public weeping, and honking car horns. The

National Foundation for Infantile Paralysis (later called the March of Dimes) funded the trial and purchased doses of Salk vaccine for rapid distribution to first- and second-graders across the country.

There had been concern about polio vaccine distribution also, but it largely stemmed from the enormous number of individuals who wanted immediate access to this new miracle of science. New York City instituted a mandatory registry of polio vaccine doses to deter corruption. The city's mayor contacted President Dwight D. Eisenhower to request direct federal intervention in vaccine allocation. Eisenhower, along with the American Medical Association, opposed "socialized medicine," including central governmental involvement in vaccine distribution, but political pressure continued, culminating in the passage of the Poliomyelitis Vaccination Assistance Act of 1955, with $25 million earmarked for vaccine purchases and $5 million for immunization programs. Measles vaccine had no such central support, at least initially.

From the beginning, the announcement of measles vaccine availability was plagued by contradictions. News reports and television specials stoked excitement about yet another triumph of modern science over an ancient infectious disease, but accompanying messages came with an ambivalent, nearly apologetic tone. News reports emphasized the local and global importance of measles, but there was no clear plan for the purchase and distribution of vaccine. Two vaccines had been approved, but the one with fewer side effects was said to be inferior. Inconsistent messages, unclear availability, and limited public understanding of why one might want to vaccinate a child against measles in the first place together presented substantial challenges that hindered the availability and success of the measles vaccine in the years following its approval.

These lessons—that messages matter, that policy decisions about vaccine availability and recommendations matter, and that early confusion on the part of parents and pediatricians can prove difficult to overcome—would resonate beyond the measles vaccine in later years.

We still grapple with these issues today, and the result is low acceptance rates for some other important pediatric vaccines. COVID-19, for which pediatric vaccination rates are much lower than they should be, is one example, but the rates of uptake of other important vaccines—including those for influenza, human papillomavirus, and varicella—suffer similarly.

7.

Vaccines Don't Save Lives. Vaccinations Save Lives.

The title of this chapter is a quote from Dr. Walter Orenstein, former director of the United States Immunization Program, former assistant surgeon general of the United States, veteran of smallpox eradication, and a man who calls measles "the disease I love to hate." The point of the saying isn't that vaccines don't work, it's that they can work only if we use them.

I took care of a critically ill two-year-old in an intensive care unit several years ago. She had contracted chickenpox from an older cousin at a family gathering and had developed a fast-moving skin infection from group A *Streptococcus* (the same kind of bacteria that causes strep throat). That infection had spread from her skin to her blood and over the course of hours had become life-threatening. Doctors at our hospital had to support her breathing with a ventilator and to use intravenous antibiotics to get the infection under control. She spent days in the ICU. The issue here wasn't that we didn't have a vaccine to prevent infection with the virus that causes chickenpox (we did). It wasn't that she wasn't old enough to have gotten the vaccine or that it wasn't recommended for her (it was). And it wasn't that the vaccine was unaffordable (it would have been available without charge). The issue was that we, the big "we" who owed it to this child to protect her—her parents, her pediatrician, the parents and pediatrician of the older child who gave her the chickenpox—didn't finish the job. We didn't use the tools that we had to make sure that this situation wouldn't happen.

There were innumerable scientific barriers to overcome to make a safe, effective measles vaccine. Only then could the next set of challenges—safe production, licensing, distribution—be addressed. If after all of that, families couldn't get access to the vaccine or didn't want to give it to their kids, it would all come to nothing. Children would still die from or be injured by measles. Developing a vaccine is essential to solving the problem, but it isn't sufficient. The vaccine has to make it into the arms of the people who need it.

The intertwined challenges of vaccine availability and access are still relevant today. The development of the COVID-19 vaccine was a marvel of efficiency. Our ability to get it to people and effectively explain its importance, less than marvelous. The decisions that we make, the way that we communicate, the messages that come from vaccine developers, governmental leaders, public health professionals, family doctors, family members, friends—all affect how well a vaccine prevents disease in the real world. If it's done well, we get smallpox eradication and polio control. Done poorly, we get confusion, fear, and preventable suffering.

The early years after measles vaccine approval were a vivid demonstration of both its potential to prevent a serious disease and its vulnerability to political missteps, poor communication, and inequitable access.

On January 11, 1962, President John F. Kennedy delivered a State of the Union address, covering a myriad of topics—income tax cuts to battle recession, the proposed Trade Expansion Act, raises for federal workers. On the twenty-fourth page of his personal notes, Kennedy proposed a new national vaccine program. "To take advantage of modern vaccination achievements, I am proposing a mass immunization program, aimed at the virtual elimination of such ancient enemies of our children as polio, diphtheria, and tetanus." In Kennedy's copy, he took care to emphasize the importance of this endeavor to the country—

VACCINES DON'T SAVE LIVES. VACCINATIONS SAVE LIVES.

"mass immunization program" and "virtual elimination" were boldly underlined, and the period at the end of the sentence had been turned into an exclamation point.

Kennedy invoked the substantial national pride in vaccine development, culminating in the ongoing successes of polio immunization. He also leveraged the popularity of initiatives focused on child health and development, as his announcement immediately followed a discussion of hot lunches and fresh milk for schoolchildren. He had assembled widespread support for an initiative to promote childhood immunization, with the understanding that the role of the federal government would be to provide funding and expertise to support state and local programs rather than to institute a nationwide system for vaccine delivery. Even within these constraints, the plan would establish a role for the federal government in vaccine policy in a way that reached far beyond the limited programs of Eisenhower's Poliomyelitis Vaccination Assistance Act. This aspect of the Kennedy plan promised substantial benefits, such as the ability to standardize immunization strategies across states and negotiate pricing with manufacturers, while still leaving control in the hands of the states. In addition, there would be potential buffering power, allowing centralized support for immunization to continue even in times of local budgetary challenges that might otherwise divert such resources.

By May, that line in Kennedy's speech had become a draft bill. The version debated on the floor of the House of Representatives focused on the three diseases that Kennedy named in the State of the Union address, but it also deliberately left the door open for inclusion of measles vaccine, which was rapidly approaching approval. Language from the draft specifically included measles as an example of how the law could be expanded in the near future: "the draft bill also authorizes similar aid for intensive programs directed against other serious infectious diseases, such as measles, for which effective preventive agents may become available."

The Vaccination Assistance Act (VAA) of 1962 was signed into law on October 23, and grants to state and local public health departments began the following June. The VAA's passage represented a successful expansion of the federal government's role in immunization policy, which would also require formalization of the way that immunizations were evaluated and recommended for use. Some vaccine recommendations came from the Committee on Immunization Procedures (later called the Committee on Infectious Diseases) of the American Academy of Pediatrics (AAP), in their "Red Book," first published in 1938. The AAP was (and is) not a federal agency, putting their recommendations outside the government's purview. The surgeon general evaluated whether vaccines should be licensed, calling upon ad hoc expert committees as needed (as he had for polio vaccine). CDC personnel tracked diseases and could make recommendations relevant to vaccines from time to time, but ongoing federal involvement in setting immunization policy, as envisioned by the VAA, would require a more durable and centralized approach. In response, the surgeon general chartered an Advisory Committee on Immunization Practices (ACIP) in 1964—an eight-member group chaired by the CDC director—to "concern itself with immunization schedules, dosages, and routes, and indications and contraindications" as well as identification of priority groups for specific immunizations.

The measles vaccine became available in 1963, but its actual use was limited in the years that followed. In the early post-approval period, there were psychological, financial, and logistical barriers to getting measles vaccines from company factories into children's arms. Despite public health officials' emphasis on the medical importance of measles, many American parents still considered it more nuisance than threat. Compounding this issue, in the absence of VAA support, the cost of measles vaccination averaged around $10 (the equivalent of nearly $100 in mid-2023 dollars), a prohibitive cost for many families, particularly those with multiple children. In addition, the side effects of the

VACCINES DON'T SAVE LIVES. VACCINATIONS SAVE LIVES.

live-attenuated vaccine, even when given in combination with gamma globulin, may have represented more of a deterrent to low-income families, for whom a day or two of missed school could represent a substantial challenge.

In June 1964, *The Wall Street Journal* reported that Merck had delivered 5 million doses of their measles vaccine, which meant that about 20 million children under age ten remained unprotected. Facing parental apathy and an absence of federal support, the road to measles control in the United States seemed likely to be long and difficult. In an attempt to increase both awareness and demand, Merck and the U.S. Public Health Service coproduced a twenty-minute film called *Mission, Measles: The Story of a Vaccine*, which touted the scientific achievements of Enders and his colleagues and again showcased the West Africa vaccination campaign. This strategy did nothing to help families who could not access the vaccine because of cost. Likewise, Merck's magazine advertisements promoting the measles vaccine to middle-class and wealthy mothers did not expand access meaningfully.

The predictable result of this situation was that the vaccine failed to eliminate measles from the United States. Because the children of wealthy families had preferential access to the measles vaccine due to its cost, its approval had the perverse effect of decreasing the overall number of measles cases (albeit not nearly as much as originally anticipated) but increasing the relative burden of disease in poor communities. Measles had always taken a steeper toll on the lives of poor children, who got the disease younger and had higher complication rates. This was true in London and Glasgow decades earlier, and it remained true in the United States in the 1960s. Prior to the vaccine, though measles was experienced differently by rich and poor, it could not be completely ignored by either group—it was a universal experience. Now, with protection reserved only for those who could afford to purchase the vaccine, measles was poised to become a disease exclusively of the poor.

As the VAA neared the end of its initial three-year authorization, the Great Society programs of President Lyndon B. Johnson were underway. Consistent with the ethos of that time, which valued the role of the federal government in addressing issues of health and inequality, the 1965 VAA renewal added measles to the list of included immunizations and also expanded the federal program's involvement in community immunization programs. Making measles vaccine eligible for VAA financing increased access to it dramatically. Millions more children were immunized, and, by the end of 1966, U.S. cases fell to about 60 percent of their pre-vaccine levels. Substantial disparities remained, but CDC director David Sencer and his colleagues felt encouraged by these data as well as by other studies that suggested that population-level protection from measles outbreaks might be achievable with vaccination rates well below 100 percent. Sencer announced a program calling for the eradication of measles in the United States in 1967—a bold goal given the incomplete buy-in from American families and the substantial problems that the distribution of measles vaccine still faced.

Sencer's plan, unveiled in November 1966, was based on four overlapping strategies for measles control: routine immunization of infants; immunization of susceptible children (those without a history of either vaccination or measles disease) at the time of school entry; improved case surveillance; and rapid responses to measles outbreaks with what he called "crash immunization programs." Measles epidemics, Sencer said, "should no longer be tolerated in the United States." President Johnson strongly supported the plan, proclaiming: "Measles, so familiar in our youth, can be dealt a final blow this fall if all the children in kindergarten and first and second grades, who are not already protected, are vaccinated."

It is worth taking a moment to discuss terminology. Much of the writing about measles control from the 1960s and following decades uses the words *eradicate* or *eradication* in describing the goals of programs like the Sencer initiative. This approach is not consistent with modern usage, which defines *eradication* as complete elimination of an

VACCINES DON'T SAVE LIVES. VACCINATIONS SAVE LIVES.

infectious agent from the entire world—meaning that further treatment or prevention is not necessary. In contrast, *elimination* of an infection refers to the removal of an infectious agent from a defined geographic area, with reintroduction remaining a threat. With regard to the measles control programs of the 1960s, definitions of eradication are not generally supplied, but it is clear in essentially all cases that what is being discussed is elimination of measles from a defined area (usually the United States) rather than global eradication. There had been little organized effort to expand measles immunization globally, and some countries had substantial misgivings about the vaccine. Routine measles immunization was not recommended until 1968 in the United Kingdom and France and the early 1970s in some other European nations. Measles vaccine was not available in many areas of the world until even later, except as part of sporadic relief efforts.

One disease was well on its way to eradication. The Global Smallpox Eradication Program had been initiated by the World Health Organization (WHO) in 1959, but progress had been slow. Intensified attempts at control, particularly in some African nations, began in the mid-1960s, and some of these efforts also included measles vaccination. Dr. William Foege, who would later design the ring vaccination strategy that would lead to successful smallpox eradication, recognized the challenges of addressing local smallpox outbreaks and simultaneously attempting to provide mass immunization against both smallpox and measles. (Ring vaccination involved identification of cases of smallpox and then vaccinating the case's contacts as well as the contacts of those contacts.) However, he also recognized the crucial role of measles vaccination in West and Central Africa at that time, calling measles "a horrendous disease . . . feared by parents and health officers." Any alteration to the smallpox strategy would have to pay attention to measles control as well. Its importance likely outweighed that of smallpox in that time and place. The last case of smallpox in the twenty-country region of West and Central Africa occurred in 1970, a crucial step on the road to global eradication, but the measles piece was left unfinished. The program ceased before local capacity to continue immunization had been ensured, leading to local resurgences of measles and the erosion of hard-won progress.

A similar pattern of initial success followed by neglect and then resurgence occurred in the United States. The 1967 measles eradication/elimination goal generated increased funding and public awareness. Billboards and newspaper articles reminded parents to get their children vaccinated, and a weeklong series of *Peanuts* cartoons made sure that a younger audience didn't miss the message. A *New York Times* headline proclaimed, "Measles Have Just About Had It," and the accompanying article included a graph demonstrating the greatly decreased case rates for the current season compared to prior years.

As a result of these intensified efforts, which included delivery of nearly 20 million vaccine doses, reported measles cases dropped from more than 500,000 per year in the pre-vaccine period to a record low of 22,231 in 1968. This decrease occurred despite enhanced surveillance for measles cases, which, if anything, could have masked progress by uncovering cases that might have previously gone unreported. Unfortunately, the early success was not maintained. Cases rebounded in 1969, and by July of that year, the newsletter of the American Medical Association declared that measles eradication had stalled and quoted a CDC epidemiologist who said that measles control would likely take several more years. In the thinly veiled racist language of the day, used by press and health professionals alike, the blame for the campaign's failure fell upon families in the "hard-core ghetto" that were thought to be "difficult-to-reach."

U.S. cities, mainly those on the East Coast, generally had lower immunization rates and higher case numbers than less densely populated areas. Public vaccination clinics were not often set up with working families in mind. Limited clinic days and hours blunted access. In addition, some factors not specific to cities hamstrung national measles immunization. While some states had successfully instituted immunization reminder services for new parents, most had not yet enacted measles immunization requirements for school entry. In addition, training and maintaining the requisite workforce for improved measles surveillance had proved difficult for many states. Measles vaccination infrastructure was underfunded, undertrained, and underavailable.

The ultimate failure of the 1967 eradication push was multifactorial. In his 1971 postmortem on the initiative, J. L. Conrad from the CDC's Immunization Branch pointed to specific objectives that were not achieved (vaccination rates of one-year-olds, institution of school mandates), but laid the blame squarely on elimination of federal funding for community immunization programs upon VAA expiration at the end of June 1969. The Nixon administration, with a substantially different view of the role of government in helping needy individuals, had used similar tactics to hinder other legislation important for the health of children from low-income families. These included the Lead-Based Paint Poisoning Prevention Act and the Emergency Health Personnel Act of 1970, both of which were enacted but languished, unfunded or underfunded by the federal budget.

Another proximate factor in the measles program's failure came unexpectedly from a different vaccine's success. Hilleman's group at Merck developed and tested an anxiously awaited vaccine against rubella (sometimes called "German measles"), which was licensed in 1969. Rubella vaccines from other companies followed in subsequent months. Rubella is generally a benign disease of childhood, causing a few days of fever and rash—and none of the complications for which measles is known. However, rubella exposure during pregnancy can lead to congenital infection (infection of the fetus during development) due to the virus crossing the placenta. Congenital rubella infection can result in miscarriage or in children being born with congenital rubella syndrome, which can include heart defects, cataracts, and neurological devastation. The massive U.S. rubella outbreak of 1964 that resulted in tens of thousands of newborns with congenital rubella syndrome still loomed large in the country's consciousness, and, given the usual pattern of five-to-nine-year cycles of rubella outbreaks, time was of the essence for development and distribution of that vaccine.

The new availability of rubella vaccines was a boon for the country, but given shifting fiscal strategies in Washington, DC, this factor likely exacerbated the damage to the measles immunization effort. Prioritization of the emerging vaccines for rubella over those for measles contributed directly to the loss of funding and the subsequent failure of the 1967 measles program. In 1971, Merck's combined measles-mumps-rubella vaccine was licensed, addressing the specific issue of choosing among individual vaccines and also decreasing the number of injections that an individual child would need, but by that time the first United States push for measles control was irreparably damaged. Some federal support eventually returned to measles immunization, but erratic funding is not a recipe for sustained public health success. Disease rates fluctuated over the next several years, ebbing and flowing with the federal government's willingness to help states provide the means of prevention.

Despite the frustrations of physicians and public health officials at the failure to eliminate measles, some important lessons came out of this period. First, despite the challenges of mass immunization in crowded cities, the overall national reduction of cases by more than 95 percent had been rapid, even if it was not sustained. In addition, in some places that reduction *was* sustained. In sparsely populated Alaska, vaccination led to interruption of indigenous measles transmission for thirty-eight months, from 1973 to 1976. Alabama, Wyoming, and Oregon all had similar successes, each lasting more than a year. Evidently, pushes for vaccination could work—measles transmission could

VACCINES DON'T SAVE LIVES. VACCINATIONS SAVE LIVES.

be halted, perhaps permanently, if we applied dedicated and sustained effort.

In addition, requirements for measles vaccination at school entry were unevenly enacted and enforced, in part due to arguments about liberty and in part due to doubts about their effectiveness. A measles outbreak in the city of Texarkana provided strong evidence that school mandates and community immunization programs worked. As was the case for the Faroe Islands, the geography of Texarkana was key to the insights that it provided into contagion and protection. True to its name, Texarkana straddles the border between the states of Texas and Arkansas. In the 1960s, about two-thirds of the population lived in the part of the city that was in Bowie County, Texas, with the remainder residing in Miller County, Arkansas. The division did not generally affect the day-to-day workings of the city, with residents of both counties attending the same local businesses, churches, and events. However, separate public schools and public health departments were maintained on either side of the state line. Texarkana was a natural laboratory to understand how policy choices could dictate health.

In late June 1970, a five-year-old Texarkana boy who had traveled out of the area was diagnosed with measles. He represented the first recognized case in an outbreak that would last more than six months and involve more than six hundred people, mostly children. That isn't the remarkable part—measles outbreaks were becoming more frequent everywhere. What made Texarkana different is that State Line Avenue separated two jurisdictions with quite different approaches to measles vaccination. Texas had no requirement for measles vaccination prior to school entry and generally eschewed mass vaccination campaigns. Fewer than 60 percent of one- to nine-year-olds on the Texas side were immune to measles either through vaccination or prior illness. In contrast, Arkansas maintained a school mandate and had held mass immunization campaigns for preschool- and school-aged children in each of the two years prior to the outbreak. An estimated 95 percent of their one- to nine-year-olds were immune.

The result was striking. A political division, not a physical one, determined who got measles and who didn't. Of the 633 Texarkana measles cases, 606 (nearly 96 percent of the total) occurred in people who resided in the Texas portion of the city. This disparity in rates occurred despite significant contact between residents from the two sides. The messages were clear—vaccination had protected children who happened to reside on the Arkansas side of town, and community campaigns and school mandates were highly effective in preventing measles spread. The Texarkana story is frequently cited in public health circles and is used as a teaching case for students of epidemiology. The lesson that is sometimes missed is that in addition to showing that vaccination protects against disease, the Texarkana measles outbreak also provides a stark reminder that political decisions about funding for public health, acceptability of school mandates, and myriad other issues can have real and lasting effects on the health of populations.

Today, Texarkana's unusual geographic and political arrangement continues to instruct us about the deeply intertwined nature of politics and health. Under the Patient Protection and Affordable Care Act (ACA, also known as Obamacare), which was passed in 2010, states were required to expand Medicaid coverage to nearly all adults with incomes up to 138 percent of the federal poverty level, with coverage going into effect in 2014. A 2012 Supreme Court decision (*National Federation of Independent Business v. Sebelius*) made states' acceptance of the ACA's Medicaid expansion funds optional rather than mandatory. Arkansas accepted Medicaid expansion; Texas did not. Jonathan M. Metzl, author of *Dying of Whiteness*, has chronicled how states' political decisions, including accepting or refusing Medicaid expansion, can change the health—and even alter the life expectancy—of its citizens. It is hard to imagine a place where that line is so sharply drawn as in Texarkana. A 2023 report from Public Health Watch detailed a

VACCINES DON'T SAVE LIVES. VACCINATIONS SAVE LIVES.

"widening divide in health care access" between the two sides of Texarkana on the basis of those decisions. Despite the demographic similarities between the Texas and Arkansas sides of the city, after nine years of Medicaid expansion, the differences were stark. More non-elderly adults uninsured, more hospitalizations for life-threatening conditions such as diabetic ketoacidosis, worse access to care—these are the legacy of Texas's refusal to accept the ACA's Medicaid expansion. Just as the 1970 measles outbreak made clear, in Texarkana, living on the wrong side of State Line Avenue can be hazardous to your health.

8.

Imperfect Tools

Nothing that we use in medicine is perfect. All medications have potential side effects. All procedures have risks. No treatment works 100 percent of the time. The same goes for preventive instruments, like vaccines. Our tools are flawed, and thoughtful doctors take that into account with every patient. We spend a lot of time in medicine balancing risks and benefits—and trying to help families make decisions that do the same. When I talk to a family about a medication or a vaccine, sometimes it is clear that the parents want me to say that what I'm recommending is absolutely safe. I can't do that. I can take them through how I think about safety, weighing the risks of doing something against the risks of not doing it. Sometimes I say the words "There is no such thing as a no-risk medicine/vaccine/procedure," but I pair these with a discussion of the risks we choose to bear when we decide not to use the incredible, though imperfect, instruments of medicine.

The road to elimination of measles from the United States in the year 2000 was full of missteps, some of which had to do with the imperfections of our tools. One of those tools was the vaccine itself, which, we would learn, protected the vast majority—but not all—people who received it. Other strategies—including political and legal means of making the vaccine widely available, communication with families about the benefits of vaccination, and overcoming inequitable health systems—all affected how we got to that elimination milestone and continue to matter today as we work to maintain it. Imperfect tools are the only kind that we have, but sometimes, applied steadily and with an understanding of their limitations, they are enough.

Texarkana, the divided city, showed us that school vaccine mandates work—they increase vaccination rates and consequently prevent disease. In the 1970s, Los Angeles would demonstrate that while mandates are useful, they, too, are imperfect—they work only when they are enforced.

A measles outbreak began in Los Angeles County in October 1976. By the following spring, more than a thousand people had developed measles, and no clear end was in sight. A combination of decreased funding for immunization drives and a systemic failure to enforce school immunization requirements led to the accumulation of a susceptible school-age population, where the cases concentrated. Dr. Shirley Fannin, head of the Los Angeles County acute communicable disease control division, saw cases accelerating among unvaccinated students despite a vaccine mandate for school attendance. She expressed her frustration at the lack of enforcement of the law: "Now we have kids getting sick who shouldn't be sick. The whole mess stinks!"

On March 31, 1977, Liston Witherill, director of the Los Angeles County Department of Health Services, ordered all schools, public and private, to bar students without proof of vaccination or immunity from attending classes. Additional vaccines and supplies arrived from the state health department and CDC, and vaccination clinics were set up throughout the county. Nevertheless, many parents were slow to respond until the May 2 deadline was imminent. On May 1, the *Los Angeles Times* reported that 35,000 schoolchildren had received "last-minute" measles vaccination. The district barred about 40,000 students, but the majority of these were vaccinated in the subsequent days and were back in classes shortly thereafter. By the end of May, 3,600 students remained on the exclusion list. The outbreak had peaked and was waning by then, but the overall toll was substantial, including more than 2,700 measles cases, at least two deaths, and numerous complications such as encephalitis and pneumonia.

The 1976–1977 Los Angeles County outbreak illustrated both the importance of maintaining ongoing immunization programs and the fact that school mandates could effectively modify parental behavior. In marked contrast to future outbreaks in the same area, only about 2 percent of families had invoked personal belief (including religious) or medical exemptions to vaccination. Many of the last-minute vaccinators described logistical issues, including lack of awareness of the requirement or of where to get vaccinated, rather than vaccine hesitancy, as the reason for delay. This is a point with modern resonance—the details that make vaccination feasible in busy people's lives have a real effect on getting shots into arms. Likewise, policy solutions, including school mandates, work, but only if they are enforced.

Both the Los Angeles measles outbreak and the public health response that led to its termination were emblematic of broader national trends. Measles case rates rose dramatically in the 1970s as a result of years of Republican presidential administrations—Nixon, followed by Ford—that underfunded federal programs designed to provide support for immunization. Even among states with laws that required children to be immunized against measles and other diseases prior to school entry, monitoring and enforcement remained lackluster. For California, it had taken a major epidemic to catalyze such action, and in many states with slowly rising cases but no major outbreak yet, the political will was not yet there.

Thus Jimmy Carter's inauguration in January 1977 came amid stagnant childhood immunization rates and outbreaks of vaccine-preventable infections that were far worse than they had been when the prior Democratic president (Johnson) left office in 1969. Carter's view of the federal government's role in health policy differed greatly from those of his predecessors. As a candidate, he had expressed support for the idea of national health insurance, and there was hope among

Democrats in Congress that their new president would value vaccination programs. In the first months of his administration, Carter propelled the creation of significant immunization policies that would reverse the trends of the prior eight years and even make the possibility of measles elimination seem achievable once again.

The first steps toward the new immunization program came not from Carter himself but from a dinner conversation. Betty Bumpers and Rosalynn Carter had met at a time when each was the first lady of a southern state. Bumpers's husband, Dale, served as governor of Arkansas from 1971 to 1975. When her husband was elected, Mrs. Bumpers expressed disinterest in the inner workings of government, hoping that not much would be required of her as first lady. Despite her initial ambivalence, Mrs. Bumpers found a policy issue about which she was passionate. Driven by the abysmal immunization rates in Arkansas—among the lowest in the nation when her husband took office—she became an advocate for vaccination programs. Her approach, culminating in the "Every Child by '74" campaign, was multifaceted, driven by volunteerism, and highly successful.

Bumpers's Arkansas program combined outreach—"bumper stickers, a lot of television and radio publicity, and press conferences and what have you"—with support from organizations inside and outside of the state government. Immunization drives commandeered medical associations, the state health department, the agricultural extension service, and National Guard units, once immunizing more than 300,000 Arkansas children in a single weekend. In addition, the program addressed issues of access—just as applicable to Arkansas families as to parents in Los Angeles—expanding clinic hours to accommodate working families and helping individuals find and make use of immunization services. The first lady's work benefited from her husband's position, but her passion for childhood immunization turned Dale Bumpers into a staunch supporter of the cause in his own right.

The Bumpers and Carter families had dinner together within weeks of Carter taking office, and Betty took the opportunity to engage Rosalynn on the topic of vaccines. Betty had been unable to get the attention of the Ford White House, but predictably, she found a much more receptive audience in the Carters. Within a day of the dinner, both the president and the first lady had contacted the newly installed secretary of Health, Education, and Welfare, Joseph Califano Jr., to say that childhood immunization would be a priority of the new administration. Califano and Bumpers met on February 18, and she explained her aggressive approach to the problem of lagging immunization rates. That same day, Dale Bumpers addressed the health subcommittee of the House Appropriations Committee, requesting dedicated support to address the chronic underfunding of immunization programs under the Nixon and Ford administrations. Senator Bumpers referenced the ongoing measles outbreaks—which by then were occurring in California, Michigan, Idaho, Virginia, Maryland, and other states—and argued that the sheer number of American children susceptible to vaccine-preventable diseases represented "one of the most abysmal failures in the medical history of our Nation."

On April 6, 1977, Secretary Califano announced a nationwide childhood immunization initiative. Both the wording of his announcement, which called the failure to protect children against preventable diseases "a shocking disgrace," and its structure ("a campaign involving the media, federal, state, and local officials, and private organizations") owed a substantial intellectual debt to the ideas of the Bumpers.

The initiative had two clear goals and a timeline. Califano's primary objective was to increase the national rate of childhood vaccination from approximately 60 percent to at least 90 percent. In addition, the program would develop a permanent system for vaccination of each year's new birth cohort—about 3.3 million per year at that time. The timeline indicated that both of these goals should be achieved by October 1979. Increased and stable funding from the federal government accelerated the tools of the program: education campaigns, statewide calls for volunteer help (using Arkansas's success as a model), improved surveillance of both measles cases and local immunization rates, rapid responses to outbreaks, and institution and enforcement of school vaccine mandates—the cornerstone of the strategy.

The public outreach portion of Carter's program was particularly ambitious. In addition to calling upon nongovernmental organizations like the American Red Cross and the National Parent Teacher Association as partners in encouraging immunization, the CDC developed flyers and booklets for mass distribution. A "Stay Well Card" originally developed by the Ohio Department of Health was promoted nationally, and medical organizations including the American Academy of Pediatrics developed and distributed their own materials. Closer to home, the initiative partnered with corporations. Kellogg's included cartoons about measles immunization as part of their "Cereal Serial" campaign, and in case breakfasting families had missed the point, the measles elimination logo was printed on milk cartons, along with the recommended immunization schedule. Professional sports leagues ran public service announcements during games, and Califano appealed directly to popular advice columnists to promote immunization to their audiences. Posters featuring characters from *Star Wars* and skits on the popular children's television show *Romper Room* helped get messages to children and their parents.

Much later, during the COVID-19 pandemic, as debates about the importance of COVID-19 infection for children's health and the need for childhood vaccination against this new disease intensified, I purchased one of those CDC posters from the late 1970s—the one featuring R2-D2 and C-3PO, the *Star Wars* droids, asking "Parents of Earth, Are Your Children Fully Immunized?"—as a reminder directed at me and anyone who might see it in the background during an interview or a videoconference that our messages about vaccines needed to be creative, clear, and accessible to families. The droids also remind me that

IMPERFECT TOOLS

the problems that we face with vaccine acceptance today—access, apathy, doubt—are not new.

Early signs of the Carter initiative's success accumulated. In 1977, 4.5 million children were immunized against measles via public funding, a substantial increase over prior years. School mandates became law. Enforcement of those mandates improved, even in areas that did not have an active measles outbreak. Most important, after the concerning rise in cases over the prior few years, measles case rates dropped substantially by the end of 1977 and in the early months of 1978.

Not all aspects of the initiative were equally successful. School audits and entry requirements worked well; the programs designed to bolster immunization of younger children, less so. The link to school attendance turned out to be crucial. Immunization reminders for new parents had mixed results, and overall immunization rates in children aged one to four years stayed concerningly low.

Despite these setbacks, by early 1978 the nationwide immunization initiative appeared to be working quite well. Disease rates continued to fall; vaccination rates rose. These successes led Secretary Califano to propose what historian Elena Conis has called a "familiar-sounding goal," the elimination of measles from the United States. In October, Califano praised the initiative's progress and announced his request for several million dollars of additional funding to extend the immunization program's work through the elimination deadline of October 1, 1982.

Public health and medical audiences strongly supported the elimination goal. Surgeon General Julius Richmond considered measles elimination both an opportunity and a responsibility for the country. CDC officials claimed that "the present situation in the United States seems like the most favorable ever for the elimination of indigenous measles."

The Carter era elimination goal was not the same as that of the eradication initiatives coming before it. A more nuanced understanding of measles epidemiology and lessons learned from prior attempts led to a careful selection of words: national elimination, not global eradication. Imported measles cases and those traceable to importation within two generations of transmission were not considered "indigenous" and would not imply failure of elimination. Such cases would still, of course, represent a threat to both achieving and maintaining elimination. Clarifying definitions helped set expectations for what a postelimination period might look like and emphasize that elimination, unlike eradication, was a state that would require ongoing vigilance and investment.

By 1981, all fifty states had laws mandating immunizations for school entry, and enforcement, though still variable, had improved substantially. More comprehensive laws and better enforcement led to lower measles incidence at the state level. As vaccination rates at school entry stabilized above 95 percent, there was a growing appreciation of the role of other populations, especially preschool children, in maintaining the chain of measles transmission in the United States. Case rates continued to decrease, reaching a nadir of just under 1,500 cases in 1983, but as the proposed deadline for elimination came and went, interruption of domestic measles transmission remained elusive.

The successes of the Carter immunization initiative were many, built on a foundation of stabilized federal funding, legal mandates, and educational programs designed to increase public awareness of and desire for vaccination. The program's goals stretched beyond reaching a certain vaccination coverage level by a certain date. Millions of American children were born every year, and in order to continue to keep measles at bay, to say nothing of making progress toward its eventual elimination, there needed to be a plan to get those kids immunized. The long game of immunization policy depended on continued public perception of scientists, physicians, public health officials, pharmaceutical companies, and the government itself as trustworthy.

Unfortunately, the yearly advances of the initiative—fewer cases, more kids vaccinated—were short-lived, and by the end of the 1980s, the United States faced yet another major resurgence of measles. This loss of progress followed a similar trajectory to that of the Kennedy/Johnson Vaccination Assistance Act, which had fallen victim to slashed funding with the changing of administrations. The programs built under Carter's initiative continued, but by the late 1980s, a combination of stagnant funding in the face of increased costs, scientific challenges in containing measles virus over the long term, and faltering public trust in institutions all contributed to an erosion of the prior gains.

Though immunization programs continued in the United States and vaccination rates among school-age children remained high, measles persisted. It spread among preschool children, a group for whom Carter's policies had proven less effective. Politicians seeking to amplify success stories could focus on the 95 percent of school-age children who had gotten their shots, but through its sheer persistence in the population, measles showed us that there was plenty of work remaining.

Young children were not the only ones getting sick in these outbreaks. Dishearteningly, it became clear that even in the setting of high vaccination rates, a small percentage of vaccinated children remained susceptible to measles. While the measles vaccine, whether given alone or as part of the MMR combination vaccine, is highly effective, it is not 100 percent effective. About 5 percent of children who receive a single dose of a measles-containing vaccine will have "primary vaccine failure," meaning that they do not develop enough of an immune response to confer protection. That means that even if every child over age five had received the measles vaccine, there would still be susceptible children in essentially every school. For that reason, effective measles control depends on increasing vaccination rates across all ages in order to surround susceptible individuals with a large, protected population. Failure to effectively immunize the one- to four-year-olds played an important role in the ongoing endemic measles spread in the 1980s.

And then there was the issue of trust. The success of the Salk polio vaccine trial in 1955 represented a high-water mark for public enthusiasm for and trust in science and medicine. In the 1970s, Carter's programs still relied on that accumulated goodwill as they asked in ever more creative ways for parents to immunize their children. Politicians passing new vaccine mandates or enforcing old ones with new vigor believed that their constituents would not vote them out of office for doing so—that citizens and their representatives shared a positive view of the vision of improved child health and the chosen means of getting to that goal. Each of these factors required an article of faith on the part of the public, an assumption that U.S. institutions and policies were working in the best interests of their families.

A lot had happened to shake that trust between the 1950s and the 1980s. Crumbling faith in governmental decision-making came into full view with the war in Vietnam and the Watergate debacle. The decreasing status of the medical profession was a slower burn, but one that was certainly underway. In the 1970s, there were increasing calls to treat health care as a right rather than something to be purchased by the few. This sentiment was combined with a growing recognition that skyrocketing medical costs were likely related to a misalignment of incentives between physicians and patients. Institutional and professional beneficence were no longer assumed by patients and their families.

This shift in attitude was not solely about medical care. Over the same period, biomedical research received scrutiny in new and quite public ways. The close of World War II and the Nuremberg Trials revealed atrocities committed by Nazi doctors in the name of science, but such horrors were not generally thought to be relevant outside of the Holocaust. The unmasking of the Tuskegee experiments, in which Black men with syphilis were studied as their disease progressed and was transmitted to others even once curative treatment was available, shattered that misconception. Even before that revelation, there were other warning signs of unethical conduct in American biomedical science, some of which involved studies of infectious diseases and vaccines in children. Together, shifting attitudes about the medical establishment and horrific examples of malfeasance undermined public faith and ultimately contributed to an erosion of confidence in vaccines.

One highly publicized case involved a central figure in measles research, Dr. Saul Krugman, though his experiments that aroused the most public outrage involved a different disease. They were part of a secretive national program of intentional infection of human subjects with hepatitis. Krugman, a pediatrician and infectious diseases specialist at Bellevue Hospital in New York City, had participated in early trials of Enders's measles vaccine. In the early stages of measles vaccine development, Krugman was eager to use his established population of research subjects—institutionalized children residing at the Willowbrook State School on Staten Island, New York—in vaccine testing. By the early 1960s, when the measles trials began, Krugman already had logged years of experience conducting research at Willowbrook. Funded by the U.S. government through the Armed Forces Epidemiological Board, Krugman and his collaborators admitted scores of children to a designated research unit at the school. There, in a quest to better understand infectious hepatitis, they carried out experiments that included deliberate infection of children—either by having them drink "milkshakes" contaminated with feces from patients with infectious hepatitis (now called hepatitis A) or, later, by injecting them with infected serum from patients with serum hepatitis (hepatitis B). Scientific papers reporting the results of his work were published in the most prestigious medical journals, fueling Krugman's rapid rise from a junior researcher to the helm of the Department of Pediatrics at New York University. Eventually, this work would bring him some of the highest accolades in academic medicine. However, his accomplishments were built on the backs of disabled institutionalized children who were placed at risk intentionally—some without any hope of benefiting from the experiments. Once the details of the hepatitis studies became widely known, controversy over the Willowbrook experiments would follow Krugman for the remainder of his career. In 1972, more than 150 people protested in Atlantic City as he received an award from the American College of Physicians. Given Krugman's considerable visibility in the vaccine field, uproar over his work provided a direct link between vaccine studies and public outrage about unethical research.

In addition to this crisis of credibility for medical institutions in general and for one of the most prominent scientists in the measles vaccine field in particular, there was also a broad change in the tenor of public discussions about vaccine safety. Against the backdrop of growing calls for regulation of both research and medical care, several high-profile legal cases addressed the issues of how risks and benefits of vaccines should be judged and who should have the power to make decisions about how to balance those factors. One of the earliest of these took place in the early, jubilant days of the polio vaccine rollout.

The joy of the initial announcement of the Salk vaccine trial—the vaccine worked!—gave way to the real logistical challenges of safely producing and distributing millions of doses to the nation's children. Rather than entrusting a single company with this daunting task, the vaccine was licensed to five companies simultaneously, each of which would manufacture, assess for quality, and distribute their own lots of polio vaccine. One of these companies was Cutter Laboratories, based in Berkeley, California. Preparation of the Salk vaccine involved growing large amounts of virulent poliovirus and then inactivating it with the chemical formaldehyde. Tragically, inadequate inactivation procedures and lax oversight at Cutter led to some children receiving injections that still contained the active virus, leading to paralysis. The Cutter Incident, as it came to be called, has been chronicled in detail by vaccinologist Paul Offit. Offit argued that the jurors in the first resulting lawsuit, *Gottsdanker v. Cutter Laboratories*, "reluctantly opened Pandora's Box" by finding that Cutter was financially liable for Anne

Gottsdanker's paralysis because they had breached an "implied warranty," despite not finding the company negligent in their production of polio vaccine under the standards of the time. The downstream effects of that verdict included an explosion of lawsuits against vaccine manufacturers in subsequent decades. Freed of the need to establish fault on the part of companies, plaintiffs sought larger and larger verdicts. This trend in turn led to rapid and significant jumps in vaccine prices and, eventually, to an exodus of companies from the vaccine market.

Lawsuits involving polio vaccines also led to new requirements to inform families about potential risks of vaccination, even when those risks were extremely rare. In 1970, eight-month-old Anita Reyes received the Sabin oral polio vaccine at a clinic in Mission, Texas. Two weeks later, she developed paralytic polio. Rarely, the attenuated viruses in the oral polio vaccine can develop mutations that result in those weakened strains regaining the capacity to cause disease, though not to cause outbreaks. Such cases of vaccine-associated paralytic polio (VAPP) occur less than once for every two million doses of oral polio vaccine, but for the Reyes family, the important thing was that Anita had been well and then, following the vaccine, she was sick. The family alleged that even though VAPP was a known adverse effect of the live-attenuated vaccine, they had not been warned of this possibility by the clinic nurse who gave Anita the vaccine. In 1974, a federal court ruled that the vaccine manufacturer had a responsibility to provide sufficient information about risks directly to families so that they could make informed decisions.

The same period brought growing concerns about combination vaccines that protected against diphtheria, tetanus, and pertussis. The idea that DTP vaccines might have significant and underappreciated dangers was not a new one. In 1973, Dr. John Wilson proposed to the members of the Royal Society of Medicine in London that the pertussis component of the DTP vaccine caused brain damage in children.

Wilson was not a fringe figure; he was a respected pediatric neurologist whose concerns were taken seriously. His findings were published in a mainstream medical journal, despite the flimsiness of his actual data. Months later, he appeared on a popular news program with the same message, igniting a media frenzy. Frightened parents and physicians backed away from DTP, leaving large numbers of children unprotected and setting the stage for the nationwide pertussis outbreak that followed.

Concern about DTP spread to the United States. In April 1982, a program called *DPT: Vaccine Roulette* aired on a Washington, DC, television station. This show provided emotionally wrenching testimony from parents of neurologically disabled children as well as purported experts who linked the children's suffering to DTP, particularly to the pertussis component of the vaccine. Though its major points about long-term damage from pertussis vaccine would all subsequently be disproven, this emotionally charged broadcast has been said to have heralded the birth of "the modern American anti-vaccine movement." The version of the DTP vaccine in use at the time certainly had rare short-term side effects. Some of these, such as high fevers or staring spells, could be quite frightening, and the combined vaccine in use today uses an acellular version of the pertussis component, resulting in greatly decreased side effects. The trade-off for this decrease in reactogenicity is that the acellular vaccine provides less durable pertussis prevention than the former whole-cell preparation.

Audience reaction to *DPT: Vaccine Roulette* was swift and practically overwhelming. The station's switchboard was flooded by calls from parents. The staff put some of the families in contact with others, unwittingly catalyzing the formation of an anti-vaccine organization, Dissatisfied Parents Together, that would go on to national prominence. The program was rebroadcast widely, and excerpts were aired on a national morning show, giving conjecture and fear a much larger platform. Predictably, parents, terrified by the images of "vaccine damaged" children, refused to consent to DTP vaccines for their children.

Faced with a looming crisis in vaccine availability, the U.S. Congress intervened by passing the National Childhood Vaccine Injury Act (NCVIA) of 1986. This legislation created the National Vaccine Program to coordinate federal immunization efforts and to oversee the newly formed National Vaccine Advisory Committee (NVAC), a multidisciplinary body that would provide expert advice on vaccine research and manufacturing. In addition, the NCVIA created the National Vaccine Injury Compensation Program (NVICP), a no-fault system for compensating individuals who were potentially vaccine-injured funded by an excise tax on every vaccine dose. The final structure of NVICP was a compromise among many parties—physician groups, vaccine manufacturers, governmental agencies, and parent advocates. Everyone involved could find something to dislike in the final structure of the program. Companies opposed the fact that NVICP was not codified as an exclusive remedy for vaccine injury claims. Parents disliked that the system had to be used prior to seeking damages in other courts. Nonetheless, there was general agreement that the system was crucial to the survival of the U.S. immunization program.

NCVIA also established new reporting requirements. Vaccine manufacturers were mandated to communicate information about adverse events that were potentially vaccine-associated. In addition, physicians were required to report information about certain severe post-vaccination outcomes. In order to gather information in a standardized way and to comply with the requirements of NCVIA, the Department of Health and Human Services created the Vaccine Adverse Event Reporting System (VAERS). Overseen by both the Food and Drug Administration and the CDC, VAERS is a passive reporting system, meaning that it relies on outside people or groups to submit information rather than collecting data itself. Anyone can submit information to VAERS, and it does not reject reports based on plausibility or perceived importance. These factors mean that while the system can efficiently detect rare adverse events and help to generate hypotheses about potential relationships of adverse effects to vaccines, it cannot determine causality (i.e., VAERS data can't show whether any event is or is not the result of vaccination). VAERS casts a wide net by design, with the knowledge that most "hits" will turn out not to be truly related to vaccination once the necessary well-controlled studies are done. It, too, is an imperfect tool. VAERS helps generate ideas for downstream investigations—that's it. In addition, because reporting is voluntary, it is not possible to use VAERS data to calculate how frequently a particular adverse effect occurs. VAERS has been in continuous operation for more than three decades. During that time, it has proved useful in detecting signals of potential vaccine-associated complications, some of which have been confirmed using active data gathering. Its limitations have also been sources of confusion for the public and of exploitation by individuals who have sought to use its data to "prove" that vaccines cause harm.

Today, both NVICP and VAERS have been repurposed as tools of misinformation by the anti-vaccine movement. The existence of NVICP is purported to show collusion between the U.S. government and vaccine makers and to bolster claims that companies act without threat of liability. Analysis of VAERS reports, sometimes by amateur investigators and sometimes by people who should know better, are used to give credence to spurious reports of harm, generally without discussion of the inability of VAERS data to provide useful information about causality. This method of undermining vaccine confidence became particularly prominent during the COVID-19 pandemic, when "VAERS dumpster dives" yielded poorly designed papers (some of which were nonetheless published in peer-reviewed journals) and substantial news coverage. The result was that confusing and often incorrect information was passed along to the public, including people trying to make consequential health decisions for themselves and their children.

By the late 1980s, NVICP had averted the liability crisis that was forcing vaccine manufacturers out of the market, but economic issues continued to undermine vaccination efforts. Under the Reagan administration's austerity measures, funding for state immunization programs had stagnated or even declined. Decreasing federal support combined with vaccine price increases meant that state and local health departments had to make difficult choices, limiting vaccine purchases and cutting back on the personnel who were essential to getting those vaccines to the children who needed them. Consequently, measles case rates crept back up from their low point in 1983. This time, however, there was an important change in the populations at greatest risk. Whereas school-aged children had previously been the drivers of U.S. measles outbreaks, school mandates—a legacy of the Carter initiative—helped keep immunization levels high in that population. Thus the new waves of cases skewed both younger and older, concentrating in unvaccinated preschoolers and in older children and young adults who had received a single dose of vaccine years earlier.

As a result of these trends, within a few years the measles situation had worsened considerably in the United States. Discussions of measles elimination gave way to desperate efforts to contain out-of-control outbreaks across the country. The earliest of these involved older, highly vaccinated groups, but by the early 1990s, when national case numbers skyrocketed, those falling victim to measles were overwhelmingly young, unimmunized children belonging to poor families in marginalized communities.

On October 5, 1988, the campus newspaper at Kent State University in Ohio issued an unusual report. Six students had been diagnosed with measles over the prior two weeks. Over the course of a typical year, the student health center generally saw one or two measles cases, making this cluster noteworthy. These six were the earliest sign of a campus outbreak that would last for more than six months and involve 380 cases. Dr. Tara C. Smith, an infectious diseases epidemiologist, has revisited the Kent State measles outbreak using contemporary news reporting and university records. Smith's analysis provides a detailed look at one of the largest college outbreaks at the leading edge of a nationwide measles surge.

Like many colleges at the time, Kent State University did not have a pre-matriculation requirement for measles immunization. Based on a survey done several years before the outbreak, the student health center director estimated that one in five students lacked measles immunity, more than enough for measles to maintain a chain of transmission. Thus a new cluster of six cases on campus was a clear cause for concern. No additional measles diagnoses appeared in the days following the announcement, spurring a premature declaration of victory over the outbreak in mid-October. When subsequent cases were reported over the following weeks, it became clear that measles was unlikely to be a short-lived problem at Kent State. Complicating matters, college officials found it difficult to track down infected students, particularly those who lived off campus. As cases continued to climb in the subsequent months, the school administration put an increasingly strict series of orders in place to try to stop the spread of the virus, including mandatory vaccination of varsity athletes, club sports participants, and music groups. Programs that involved interaction between students and members of the surrounding community were suspended. Finally, the administration issued a broader ultimatum: students living in dormitories had to be vaccinated or move out. By late April, with more than 7,000 students newly vaccinated, the outbreak was over.

It is possible that a requirement for immunization against measles prior to enrollment—an entry requirement analogous to the ones in place in kindergartens across the country—might have prevented the Kent State outbreak. A later analysis demonstrated that colleges that had such requirements in place were at considerably lower risk of measles outbreaks, even in the context of a nationwide surge in cases. However, even ensuring that nearly all students were immunized prior to enrollment did not confer a guarantee of protection. Fort Lewis College in Durango, Colorado, is a different kind of cautionary tale. The events on that campus provide an example of how measles transmission can be maintained even in the setting of near-complete immunization of entering students. Over a two-month period starting in late December 1987, Fort Lewis had a surge of measles cases. As with Kent State, most transmissions took place on campus. Students living in the close quarters of a campus dormitory had triple the risk of infection of those who lived off campus. The factor that made this outbreak most remarkable was that nearly 99 percent of the student body had evidence of measles immunity—they had had to provide proof in order to enroll at Fort Lewis. However, not all of those students with evidence of immunity had necessarily been vaccinated. The school also accepted documentation of a physician-diagnosed case of measles in the past.

Of the eighty-four measles cases at Fort Lewis College, nine occurred in students who had not provided evidence of vaccination prior to enrollment. But those nine students made up only 11 percent of the total number of cases in the outbreak. That meant that nearly 90 percent of the cases at Fort Lewis occurred in students who *had* been vaccinated against measles. Did that mean that the measles vaccine didn't work?

In fact, the data from Fort Lewis suggested exactly the opposite. Given that the school population was overwhelmingly vaccinated, it was possible to draw a few conclusions. First, among the students at highest risk (those living in campus dormitories), the measles attack rate (i.e., the number of cases divided by the size of the population) was more than thirteen times higher for unvaccinated students than for their counterparts who had been appropriately vaccinated. This meant that for an individual student, measles vaccination was more than 90 percent effective at preventing infection during the outbreak. Most students who got measles during the outbreak were vaccinated because the vast majority of the student body was vaccinated, but a much smaller percentage of the vaccinated group was infected compared to the unvaccinated group.

This kind of epidemiologic optical illusion is a common cause of confusion that would become ever more important in later decades. It would take on particular poignancy during the COVID-19 pandemic, when outbreaks in highly vaccinated populations were held up as evidence that the vaccine "didn't work." This point is important, so I'm going to say it again a little differently. In a highly vaccinated population, with a vaccine that is anything less than 100 percent effective, we *expect* to see more cases of disease in the vaccinated population than the unvaccinated population. This observation is driven by the fact that the vaccinated population is much, much bigger than the unvaccinated one (and thus even a small percentage of that big population will be greater than a higher percentage of the smaller population). Rates (cases divided by population size for each group) are what matter. Whether we are discussing measles or COVID-19 or other infections with families, trying to dispel incorrect information online, or even talking to doctors, this distinction can be a minefield of confusion. The upshot is that population sizes (the denominators in these fractions) matter, and if you want to know whether a vaccine or any other prevention strategy works, you can't just count cases.

The other important lesson of Fort Lewis was that even though a single dose of MMR worked well, providing protection to more than 90 percent of those who received it, that low rate of primary vaccine failure left enough individual students susceptible for measles to gain a foothold and spread throughout the campus. For both individuals and the community as a whole, the more vaccinated students the better—as outbreaks would be shorter, less severe, and less likely to get started in the first place—but measles remained a risk. Outbreaks among highly vaccinated populations like Fort Lewis prompted recommendations to add a second dose of MMR vaccine to the routine vaccine schedule, a change that would further reduce, though still not to zero, the number of unprotected individuals.

The city of Chicago provides a third example of the character of measles outbreaks, one that represents the later, more severe phase of that multiyear surge. On February 14, 1989, a young woman who worked in an office in the Sears Tower went to a local hospital because she had vomiting and diarrhea. Two days later, she developed the classic rash of measles. None of her friends, family, or coworkers had symptoms of measles, and she had not come into contact with anyone from Bradley University in Peoria, about 150 miles southwest of the city. Public health officials had been monitoring the early stages of a measles outbreak at Bradley. The first case on campus had been diagnosed about two weeks before the woman's illness. In mid-March, two students from Bradley returned to their homes in Chicago for spring break. Neither had symptoms at the time that they left school, but both were diagnosed with measles shortly after their arrival in Chicago. Both students had been vaccinated against measles in childhood, as was the case for nearly all the students in the Bradley outbreak. These three cases—the office worker and the two students—were among the first in a Chicago measles outbreak that started in 1989 and lasted through 1990.

Several aspects of the Chicago measles outbreak were remarkable. One was its duration—the city would still be counting cases a full year after those three individuals were diagnosed. Another was the sheer number of people affected. By July, the Chicago Department of Health had confirmed 262 cases. By the end of August, the total was over a thousand. Intensive public health measures were implemented—lowering the vaccine eligibility age to under a year, expanding access to vaccination clinics, conducting publicity blitzes, providing vaccines in pediatric emergency departments as an adjunct to doctors' offices and clinics, and sending door-to-door vaccination teams into communities with high case rates. Despite these efforts, by the end of 1989, the number of cases had more than doubled again—more than 2,200 cases, with 755 hospitalizations and 8 deaths (7 of whom were children). The per capita rate of measles in Chicago children was more than ten times the rate in the United States as a whole.

It was impossible to miss which Chicago communities were hardest hit and whose children were getting sick. Three-quarters of the measles cases were children under age five; nearly 95 percent were classified by public health officials as either Black or Hispanic. Three-quarters of the cases were unvaccinated, and a survey of public school records revealed a massive disparity in on-time measles vaccination rates (by age two) between predominantly white (80 percent) and predominantly Black and Hispanic (approximately 50 percent) schools. Neither rate was optimal, but the 30-percentage-point difference was damning and likely represented a major factor in the disparate impact of measles on Chicago communities.

The Kent State and Fort Lewis campus outbreaks and the struggles with measles in Chicago were emblematic of measles epidemiology in the United States during the late 1980s and early 1990s. From 1989 to 1991, the United States experienced its largest measles outbreaks in more than a decade. The fundamental problems—undervaccination of young children, particularly those from marginalized communities in large cities, and unrecognized susceptibility in highly vaccinated populations such as college students—were present across the country. As a result, Chicago wasn't alone—there were significant measles outbreaks in New York, Houston, Dallas, San Diego, Los Angeles, Philadelphia, and many other cities. Kent State and Fort Lewis also weren't alone—other colleges had surges of measles on campus during this period. In 1989, more than 18,000 measles cases were reported in the United States with 41 deaths. The next year, the situation worsened—more than 25,000 cases and 60 deaths. The disparities noted in Chicago's kids were projected on the nation as whole—about half of all cases were in kids under age five. Black and Hispanic children were affected at seven to nine times the rate of white children.

Americans were shocked and frightened by the scope of the outbreaks. With the success of vaccination and widespread school mandates, the country had begun to get used to life without measles. Newspaper headlines about spiking cases and children dying of a preventable disease were disconcerting, and people wanted answers. Clearly unrecognized problems with the vaccination system in the United States were at the core of the measles surge. The tools to prevent this infection, despite their imperfections, were available. So why did the spread persist? Doctors and public health officials were frustrated. It was one thing to have children die of non-preventable diseases, but seeing it happen when effective prevention was at hand is a different kind of horror.

In January 1991, as the nationwide outbreaks waned, the National Vaccine Advisory Committee (NVAC) issued a report focused on the underlying causes of the measles epidemic. Their findings were a scathing indictment of domestic vaccine policy and a warning that in the absence of major reforms, the situation was likely to worsen soon. In their words, "The principal cause for the epidemic is failure to provide vaccine to vulnerable children on schedule." The populations hardest hit by measles in 1989 and 1990 were the ones with greatest difficulty accessing vaccination. The downstream effects of cuts to immunization funding over the course of the 1980s had become visible on a grand scale. In addition, it was clear to the committee that the recent outbreaks did not represent simply a measles problem. Instead, they were "a warning flag of problems with our system of primary health care."

The NVAC members identified impediments to timely immunization access that included the cost of vaccines to families, limited clinic hours, and language barriers. They recommended several interventions to help stem the tide of measles and to attempt to prevent the coming epidemics of other vaccine-preventable diseases. It is telling that most of their recommendations focused on immunization delivery in general (increasing federal funding for immunization programs; eliminating underinsurance for vaccines; setting up minimum standards for immunization practice; institution of vaccine mandates at day care entry), with only a few measles-specific proposals (full implementation of a two-dose measles vaccine schedule to deal with the problem of primary vaccine failure; specific studies on measles epidemiology). NVAC realized that measles had exposed gaps in the U.S. immunization system. It had appeared first, but outbreaks of other vaccine-preventable diseases were likely to follow. Additional funding was released in response to the report, but the NVAC recommendations would have longer-lasting impact, as many of them would reemerge as core elements of vaccine policy for the next presidential administration.

Arkansas governor Bill Clinton placed health care policy at the center of his 1992 presidential campaign, particularly focused on its intersection with the flagging American economy. Clinton promised comprehensive reform that would expand access to care, especially childhood immunization, to those who needed it and that would also address the ever-expanding burden that medical expenses inflicted on families. As Carter had, Clinton also appreciated and repeatedly communicated the cost savings associated with childhood immunization.

Following his election, Clinton's proposed Comprehensive Child Immunization Act took seriously the needs that NVAC had identified. Carter-era school-based vaccine mandates were already successful at ensuring coverage in those older children, so a major focus of the Clinton program had to be universal immunization of preschool children—the underimmunized high-risk population that suffered most acutely in the recent measles epidemics. The Clinton administration's goal was to achieve at least 90 percent coverage for this group, which would represent a major improvement and provide a firewall to dampen or prevent further outbreaks in American cities. The initial proposal included an ambitious mechanism to reduce cost barriers to vaccination, an issue that had taken on increased relevance. Vaccine prices surged in response to DTP lawsuits and remained out of reach for some families. The Clinton plan's universal purchase component, under which the federal government would buy vaccines from manufacturers at a discount and distribute them to public health clinics and private physicians, was opposed vigorously by vaccine manufacturers, who reasoned that it would hinder innovation, as well as by some politicians, who argued that cost did not represent a major barrier to vaccination and that free vaccination for all children would amount to a taxpayer-funded giveaway to the rich and insured.

The compromise solution that passed as part of the 1993 Omnibus Budget Reconciliation Act created a new program, Vaccines for Children (VFC). VFC took the form of a Medicaid-funded entitlement program, mandating federal purchase of all vaccines recommended by the CDC's Advisory Committee on Immunization Practices. This arrangement was (and remains) unusual, as it gives a committee of experts the power to expand government entitlements rather than leaving such decisions solely in political hands. Children eligible for VFC included those who were uninsured or covered by Medicaid, those whose insurance did not cover vaccines, and those belonging to Native American or Alaska Native groups. The legislation addressed a shortcoming of prior proposals by allowing federal funding to support not just immunization purchases but some additional related costs. That change allowed states to provide salary support for immunization providers, a major item that was missing from prior arrangements and a potential solution to the personnel shortages that had plagued local programs.

Clinton's comprehensive national program did not survive the legislative process. To overcome political resistance and get the Comprehensive Child Immunization Act through Congress, major elements had to be removed or diminished. Universal vaccine purchase by the federal government was scaled back; a national immunization registry that would have standardized and simplified tracking of children's vaccines was scuttled, leaving state and local health departments to develop their own systems. Nonetheless, the program that emerged—VFC—succeeded on a massive scale. VFC was implemented formally in October 1994. Under CDC's leadership, VFC thrived, combining the twin benefits of stable federal funding and flexibility in local implementation. Immunization rates among preschool-aged children soared. A remarkable element of VFC has been its staying power, due in part to the fact that as a federal entitlement program it is not subject to fluctuations in annual appropriations. As a result, VFC continues to provide access despite substantial increases in the number of recommended vaccines as well as their prices. In its first twenty years, VFC prevented 322 million illnesses, 21 million hospitalizations, and 732,000 deaths. The disparities in immunization rates across racial, ethnic, and socioeconomic lines that were so clear in the measles outbreaks of the late 1980s and early 1990s have decreased substantially. Thanks, VFC!

VFC's early successes in boosting measles immunization rates brought the United States closer than ever before to the always elusive goal of measles elimination, which has been part of the rhetoric around measles vaccination campaigns since their inception. As the new millennium approached, the measles-specific components of the push for elimination—vaccination of preschoolers, a routine two-dose series, aggressive surveillance, and outbreak response—had clearly succeeded. Case rates were down. Disparities in access were down. Molecular epidemiologic studies using genetic data to track the spread of measles showed that, although travel-associated cases were still a challenge, no single strain of measles virus was continuously circulating in the United States. In March 2000 a CDC panel certified that the U.S. had finally—third time's a charm—eliminated measles.

It is telling (and in retrospect somewhat concerning) that this accomplishment—a public health goal discussed since the beginning of the measles vaccine programs, the subject of multiple presidential initiatives—barely registered in the national press. A decade after outbreaks swept the country, with scores of children dying of a vaccine-preventable disease, measles was no longer front-page news. Between 1997 and 1999, there had been a total of fewer than 400 measles cases reported in the country—an annual incidence of under one case per million people each year. No one was talking about it anymore. Plus, there was the added difficulty that "elimination" didn't mean a total absence of cases—there would still be travelers with measles, just not sustained spread in the population. For newspaper editors, the difference between the prior few years and measles elimination may have seemed too minor to merit coverage, if they noticed the achievement at all.

Despite the tepid domestic response, this achievement had global implications. Measles elimination in the United States did not occur in a vacuum. Encouraged by local successes in measles control with vaccines, beginning in the late 1980s, international public health bodies, including the World Health Organization, set increasingly ambitious goals for increasing measles vaccination coverage and decreasing measles-related mortality in children worldwide. By the late 1990s, prior to U.S. elimination, global measles deaths had decreased by about 85 percent. Global rates of receipt of a first dose of a measles-containing vaccine (called MCV1) had increased to about 70 percent and leveled off. Not all areas of the world benefited equally from efforts to increase availability of measles vaccines, though. Two regions, the Americas and the Western Pacific, had MCV1 rates greater than 90 percent. Fourteen countries, all of which were outside of the Western Hemisphere, had rates below 50 percent, and ten of those fourteen were in Africa, including Nigeria, the Democratic Republic of the Congo, and the Central African Republic.

The upshot of these numbers was that in the same year that the United States achieved elimination, the entire Western Hemisphere was on its way to being measles-free. Since 1994, the Pan American Health Organization (PAHO) had been working toward that explicit goal, and many countries in the region had already interrupted endemic measles transmission prior to the United States doing so. The United States had benefited from those decreased case rates in neighboring countries as it moved toward elimination. Remember that even as measles faded from the minds of Americans and public health groups quietly celebrated local successes, the world was still home to tens of millions of measles cases in children every year, with nearly 800,000 deaths directly attributable to the disease. It's just that these weren't nearby enough for most Americans to notice.

The idea that measles might be eradicated globally through vaccination, as smallpox had been in 1977, had first been discussed seriously in the early 1980s. There were no obvious reasons why it couldn't be done. Measles shared several features with smallpox that argued in favor of the possibility of eradication. Infection nearly always resulted in a readily recognizable rash (though the fact that measles was contagious prior to the onset of the rash represented a crucial difference between it and smallpox). Survivors of the infection had lifelong immunity. An effective vaccine was available. Finally, there was no known animal reservoir of infection. (As Enders had surmised, nonhuman primates can be infected by measles due to contact with humans; however, the size of primate populations in the wild is not sufficient to maintain the spread of the virus.) In 1982, William Foege, then director of the CDC, concluded that worldwide measles eradication "is worthy of our best endeavors."

Decades later, as it became clear that elimination of measles from whole continents was not just feasible but a likely reality, discussions about eradication became more concrete. In 1998, Foege appeared before a Senate subcommittee to discuss global eradication of polio (that campaign was well underway) and measles. In his statement to the committee, Foege reiterated his belief in not only the feasibility but the advisability of global eradication despite the "tremendous effort" that would be required. He saw a measles eradication campaign as a chance not only to rid the world of a major killer of children but to "make measles vaccine a tugboat to pull the entire immunization program to even greater heights." His vision was to use a measles eradication program to strengthen public health systems on a global scale.

There was a long way to go between local elimination and global eradication, and it wasn't even a sure thing that a country (like the United States) that had managed to eliminate measles would be able to maintain that status. Local elimination is a precarious state. One important challenge to maintaining elimination status was the massive global measles burden that promised continued imported cases as well as potential exposure of U.S. residents who traveled to endemic areas. Another, which we will discuss in the following chapters, was a quietly growing and maturing anti-vaccine movement within the United States and other nations. Discounted as a factor in the early 1990s, anti-vaccine sentiments were making inroads in American communities and in the emerging world of the internet. These would eventually pose an existential threat to measles and polio elimination and to the control of other diseases, including COVID-19.

9.

Amnesia

I would like to return to the measles virus itself and another way that its biology—the functions encoded in its RNA—impacts humans as both individuals and societies. Discussions of the interplay between measles and us often involve the virus's unparalleled transmissibility. However, another aspect of measles is not only blindingly cool and a major factor in its success as a driver of global mortality, but also provides an apt metaphor for how we have approached—or ignored—measles together in recent years. Measles makes us forget.

As discussed earlier, measles virus infects a human host by first entering immune system cells that express a particular receptor, SLAM. That simple molecular fact has important long-lasting consequences. Through that interaction, measles virus is able to slip past the border cells in the airways and hitch a ride to nearby lymph nodes. In those nodes, measles encounters other SLAM-expressing cells, which it then uses to disseminate itself throughout the body. But the consequences of SLAM binding go far beyond transportation. Some of the most important SLAM-expressing cells that measles virus kills are those responsible for immunologic memory.

Normally, an infection prompts an orchestrated response from a host's immune system. Innate immune cells, which detect potential pathogens by recognizing commonly occurring molecular clues, set things in motion. We are hardwired to see these patterns—a particular combination of fats and sugars that appears on the outside of bacteria triggers one kind of response; a strand of DNA that is in the wrong place inside a cell (as it might be during a viral infection) triggers another. Detecting these clues, innate cells begin the process of engulfing and inactivating infectious agents, walling off the infection site, and recruiting other cell types to participate in later phases of the response. Those subsequent steps involve targeted responses, tailored to the specific pathogen and culminating in production of memory T and B cells, which provide long-lasting defenses. Memory B cells are important for quick antibody generation when a previously encountered microbe returns; memory T cells can help kill other cells infected with a virus remembered from the past. These mechanisms are crucial for protection against a wide array of microbial challenges and are also the secret behind how vaccines provide durable protection. By inducing immunologic memory without infection, vaccines give the immune system the benefits of a prior infection without the risks.

Your repertoire of memory B and T cells functions as an encyclopedia for your immune system, with a vast number of cells ready to respond to challenges that, through either previous infection or vaccination, you have been exposed to in the past. Because measles virus can enter and destroy cells that express SLAM, including these memory cells, it effectively wipes those pages clean. It's among the dirtiest tricks I know—up there with the way that HIV ravages the body's T cells, weakening a crucial arm of our defenses. Measles erases our immune system's memory, leaving us to start over and exposing us to danger from foes we had already learned how to fight. A child who had been vaccinated against rotavirus might find themselves with rotavirus diarrhea; another who has had a respiratory infection before might experience it again as if for the first time. The overall effect is that measles leaves us vulnerable to a wide range of other infections even after the measles virus itself is long gone, a state called immune amnesia.

The mechanisms behind immune amnesia have only been worked out in recent years. However, understanding this phenomenon helps explain some peculiar observations about measles that date back more than a century. In 1908, the Viennese pediatrician Clemens von Pirquet noticed that children with measles who had previously had a positive skin test for tuberculosis (the kind of test that once someone tests positive, they should remain positive indefinitely) unexpectedly had a negative result on the same test if it was administered during or shortly after measles infection. In addition, children with measles are more susceptible to pneumonia and other potentially life-threatening infections even after they have recovered from measles. That period of increased vulnerability, called the measles shadow, can last for several years after measles infection. It is a long, dark shadow.

Immune amnesia does not occur in kids who have been vaccinated against measles, meaning that not only does vaccination protect against measles itself, but it protects indirectly against the many other infections that can lurk in the measles shadow. This fact explains a conundrum from early measles vaccine studies, in which the vaccine "overperformed," particularly in underresourced areas where deaths from pneumonia, diarrhea, and other causes are common. The phenomenon of nonspecific benefits of measles vaccination—mortality benefits going beyond what would be predicted simply from prevention of measles in a given area—was noted in several studies in different settings. Children who received measles vaccines were less likely to die not just from measles but from non-measles causes as well. We now know that those benefits derive—at least in part—from the vaccine's ability to ward off measles-induced immune amnesia, effectively shining a light on the measles shadow. Preserving memory cells that would otherwise have been lost has long-lasting beneficial effects for children.

The measles vaccine does not cause immune amnesia—it protects against it—but its success on a population level can, ironically, damage our collective memory. It's a phenomenon that we see again and again. The better that we get at using measles vaccine, the lower the case rates go. The lower the rates, the less people think about measles. Over the years, new parents may never have experienced measles themselves or known anyone who has. Politicians may wonder why money is going toward vaccination campaigns to prevent a disease that we barely see anymore. Doctors may go years without taking care of a child with measles. As we forget, it becomes harder to communicate about the risks of forgoing vaccination. It's our own kind of immune amnesia, but this time it's on a broader scale. Vaccination rates drop, cases rise, and the cycle begins again. Measles thrives when we collectively forget.

Back in 1991, when members of the National Vaccine Advisory Committee reviewed the failings of U.S. vaccine policy and proposed solutions that would become the backbone of the Clinton vaccination initiative, public resistance to vaccination was not high on their list of priorities. The committee's four answers to the rhetorical question "Why are children not being vaccinated?" were (1) missed opportunities for giving vaccines; (2) shortfalls in the health care delivery system; (3) inadequate access to care; and (4) incomplete public awareness of the importance of immunization. Vaccine hesitancy or frank anti-vaccine attitudes did not even make the list. If these ideas were considered at all, they were rolled into the idea of "incomplete awareness."

Dr. Shirley Fannin, veteran of the Los Angeles measles outbreaks in the late 1970s, echoed this attitude in her June 1991 Senate testimony. Fannin testified about her experience during the 1988–1991 measles surge in Los Angeles, describing massive case numbers (more than 6,000 reported, with 87 deaths) and costs (more than $18 million locally). She also made clear her thoughts about the underlying issues that spurred the outbreak—particularly unvaccinated young children. As a leader in an underfunded health department, Fannin had to put resources where she thought that they would make the biggest difference in the least time. Her group was unable to institute the recommended second dose of measles vaccine—they couldn't afford it—and focused solely on vaccinating the unvaccinated, particularly the "hard to reach" groups. As a result, she thought deeply about how best to get those children immunized. Her answer was to decrease barriers to immunization—clinic hours, cost, travel time—not to deal with vaccine hesitancy. In her mind, hesitancy represented a minor issue at most. Fannin acknowledged parents' ability to refuse vaccinations for their children because of their personal beliefs, but she said that exemptions were rare and were "not an impediment to our immunization program." In his testimony, Dr. Walter Orenstein, director of the U.S. Immunization Program, agreed: "[T]he bigger problem has been among people who do not object, and we have just not been able to reach until the time of school entry."

Fannin related one anecdote that foreshadowed the difficulties of the coming decades. "Now, we had a little epidemic of whooping cough in a private school among a yuppie-parent type group who rejected DPT. That is a real oddity. That is not the rule. The rule is hard to reach." During the surge of the late 1980s and early 1990s, vaccine refusal was, at least in the eyes of public health figures of the time, a much less important issue than barriers to vaccine access for families. However, Philadelphia's measles experience argues against that narrative, heralding the growing importance of vaccine refusal in the reemergence of measles.

Like many other large cities in the United States, Philadelphia endured a surge of measles in the late 1980s and early 1990s. Between October 1990 and June 1991, the city had more than 1,400 measles cases. Nine of these were fatal. The thing that made Philadelphia's experience unlike that of Chicago or Los Angeles or New York City was that vaccine refusal did play a major role in the spread of measles and the tragic deaths of several children. Consistent with national trends, the early cases in the Philadelphia outbreak occurred mainly in unimmunized preschool children. By the end of 1990, the city had seen several hundred cases, including two fatalities, and had implemented emergency measures. But early in 1991, the outbreak took on a new quality. Cases started to appear among members of two Philadelphia churches—Faith Tabernacle Church and First Century Gospel Church. Both churches taught members not to accept conventional medical care for either illness or prevention. That teaching included refusal of all vaccines.

In his book *Bad Faith*, Paul Offit tells the dramatic story of the first deaths among the children of members of these churches—a nine-year-old named Caryn, followed three days later by another nine-year-old named Monica, followed four days later by a fourteen-year-old named Linette, followed the next day by a five-year-old named Nancy. Church officials cooperated only minimally with the city health department. Families declined medical care for those who were ill and vaccination for those who had not yet been affected. A court order allowed volunteer physicians to enter the homes of families with confirmed measles cases to determine the severity of their children's illness. As a result, five children were hospitalized against their parents' wishes. Eventually, at the direction of the city's mayor, a court order authorized compulsory vaccination for preschool-aged children who had not yet been exposed. In March 1991, eight children received MMR vaccine under that order, an unprecedented action that, while it did provide protection to those individuals, happened late enough that it didn't alter the overall pace of the outbreak. By the time spread had ceased, nearly 500 cases of measles had occurred among members of the two churches, compared to 938 in the rest of the (vastly larger) Philadelphia population. This figure meant that the rate of measles was about a thousand times higher among church members than among the rest of the city, and the case fatality rate (i.e., the risk of death among those with confirmed measles) was about four times higher.

Fannin's anecdote about the "yuppie" parents refusing DTP vaccine and the tragic story of the children of the members of the Philadelphia fundamentalist churches were treated as outliers rather than the warning signs that they would prove to be. The Clinton programs dealt with logistical and funding issues. They focused on access to care. And, importantly, they worked—setting the United States on the road to measles elimination. Those programs did not address the issue that Fannin and other experts had indicated was a lesser concern—vaccine hesitancy. This failure would return to haunt the country, to threaten the gains made under the Clinton programs, and eventually to undermine the response to the COVID-19 pandemic. As measles case rates dropped in the United States and the disease slipped out of news cycles and the public's consciousness, a series of events laid the groundwork for continued growth of anti-vaccine sentiments.

On May 16, 1997, President Clinton provided a long-overdue public apology for the government's involvement in the Tuskegee syphilis experiment. Saying that "what was done cannot be undone, but we can end the silence," Clinton called the government's conduct "shameful" and the study itself "clearly racist." Reaction to the apology was generally positive, though many writers pointed out the enduring effects of Tuskegee on the health of Black Americans. Prior to Clinton's apology, Dr. Stephen B. Thomas, then the director of Emory University's Institute for Minority Health Research, directly linked Tuskegee's legacy to modern health disparities, citing reluctance to seek out testing or treatment for HIV infection, donate blood, or register as organ donors as consequences of distrust engendered by the government-funded experiment.

At times, this distrust involved vaccines. During distribution of a new vaccine during a 1996 outbreak of hepatitis A in Memphis, city residents raised concerns that Black children were being treated as "guinea pigs." Direct comparisons to Tuskegee followed. The same year, CDC director David Satcher confirmed that a 1989 CDC-funded study had tested two measles vaccines in poor Black and Latino children in Los Angeles without disclosing that one of the vaccines had not been approved for use in the United States. Satcher stated that there were important differences between that CDC mistake and Tuskegee, but the comparison came up repeatedly in the press. Saul Krugman had died in 1995, and his obituary in *The New York Times* focused on his work in the area of vaccines, particularly the measles vaccine, mentioning only briefly the controversy over the Willowbrook hepatitis experiments. However, the Tuskegee study and Krugman's Willowbrook experiments were linked in the public mind, frequently appearing together in newspaper articles about abuses in human experimentation and drawing a throughline from the horrors in Alabama to a prominent voice in modern vaccine research.

Despite enduring mistrust due in part to the memories of Tuskegee and Willowbrook, and despite the continued attempts of a small but determined anti-vaccine movement led by Barbara Loe Fisher (a founder of Dissatisfied Parents Together) to sow doubt, vaccination rates remained at historic highs in the United States in the late 1990s. For measles vaccine in particular, the combination of mandates and policy interventions to increase access led to rates above 90 percent for school-aged children—a crucial component of the successful push toward elimination.

The situation was less rosy in the United Kingdom, where two brands of MMR had been pulled from the market in 1992 due to concerns about the mumps component of the vaccines. The subsequent media panic led to a precipitous drop in MMR use and, within two years, a predictable spike in measles cases. Public health authorities campaigned to promote measles vaccination and met with considerable public resistance, particularly by members of a group called Justice, Awareness, and Basic Support (JABS), founded by Jackie Fletcher. Fletcher blamed MMR vaccination for her son's chronic neurological illnesses and had embarked on a campaign to sue the vaccine manufacturer.

In 1998, *The Lancet*, a prestigious medical journal based in the United Kingdom, published a report from Andrew Wakefield and his colleagues. Wakefield's now-notorious paper studied twelve children and promoted a radical hypothesis—that the MMR vaccination could cause previously healthy children to develop a regressive neurological condition, a kind of autism. The study was fundamentally flawed and could not have addressed the hypothesis that he proposed—the sample size was limited; there was no control group; the methods were poorly suited for the conclusions that he was trying to draw. Despite the limitations of the study, Wakefield suggested, based on his findings, that children should receive measles, mumps, and rubella components of the vaccine separately and over a long period of time in order to minimize risk. This approach would both increase the length of time that children would be vulnerable to infections without providing any known benefit and increase the number of shots that they received. His proposed solution was not tested—could not have been tested—in his study. Like so much else from Wakefield, it was fabricated out of nothing.

Journalist Brian Deer proceeded to "peel the rancid onion" of Wakefield's lies, discovering research fraud, unethical conduct, and significant undisclosed conflicts of interest (including but not limited to fabricated data, clinically unjustified invasive procedures performed on children, payments totaling hundreds of thousands of British pounds from a plaintiff's lawyer working with JABS, and patent applications for his own measles vaccine and autism treatments). Most important, Wakefield's hypothesis was incorrect, as demonstrated by a multitude of subsequent studies. As Offit has pointed out, "[Wakefield] wasn't just wrong, he was spectacularly wrong." Following a lengthy investigation, *The Lancet* retracted Wakefield's article, and he was stripped of his medical license in 2010. Deer's book, *The Doctor Who Fooled the World*, plumbs the depths of Wakefield's deceptions and the wide reach of his claims. It is a masterwork of investigative reporting that provides a cautionary tale about the potential for unethical researchers to harm the public's health.

Despite the eventual debunking of Wakefield's work, it caused considerable damage in both the short and long terms. Its effects are still being felt. In the immediate period after publication of his article, MMR vaccination rates in the UK dropped once again, leading to surges in measles cases and avoidable suffering. The longer-lasting effects have generally fallen into two categories. First, the idea of a link between vaccines and autism has proven to be "sticky." Heidi Larson, an anthropologist and the director of the Vaccine Confidence Project, attributes the longevity of Wakefield's message to a combination of its simplicity ("vaccines cause autism") and its emergence in the years just before the rise of social media platforms like Facebook and Twitter. The eminently tweetable sentiment seems to have been tailor-made for the approaching digital age.

Just as important, the *Lancet* paper provided a boost to an anti-vaccine movement that needed a figure to rally around. Wakefield took advantage of climbing rates of autism diagnoses and the lack of a clear understanding of the disease's cause, claiming that he had identified an easily understandable answer. His considerable charisma and his perceived persecution at the hands of the medical establishment have made him an ideal martyr for the cause of vaccine resistance. In 2004, after losing his job at the Royal Free Hospital, Wakefield moved to the United States, settling in Austin, Texas. There he became the executive director of the Thoughtful House Center for Children, a newly founded organization for research and treatment of autism. Wakefield has continued to deny any wrongdoing and to promote his theory that MMR vaccination causes autism. Orenstein credits Wakefield's falsified work and his communication skills with shifting the focus of the U.S. anti-vaccine movement away from DTP and other vaccines and onto MMR. He says of Wakefield, "If I didn't know better, I could buy a used car from that guy . . . He just knows how to speak." Dr. Alan Hinman, former director of the Immunization Division at the CDC and former assistant surgeon general, agrees: "I don't think that measles was as singled out before Wakefield as it has been since."

Wakefield unceasingly sought publicity. He testified at hearings, cultivated celebrities and politicians, and crusaded for anti-vaccine causes in the United States even before moving there. Indiana representative Dan Burton, chairing a 2002 hearing on vaccine safety and autism, praised Wakefield in nearly messianic terms, claiming that he "like many scientists who blazes [sic] new trails, has been attacked by his own profession . . . He and his colleagues have fought an uphill battle to continue the research that has been a lone ray of hope for parents." The celebrity Jenny McCarthy wrote the foreword for Wakefield's 2010 book *Callous Disregard*, in which she expressed strong support for both the man himself and his long-since-discredited theory.

Unlike in the UK and despite McCarthy's efforts, overall MMR vaccination rates remained high in the United States after the Wakefield publication, with rates at school entry holding at about 95 percent each year. A nationwide rate can be falsely reassuring, though. The messages that came from Wakefield, McCarthy, and others didn't have to work on a national level to do damage. Because of the prodigious transmissibility of measles, generating enough doubt to lower vaccination rates even within small communities would prove sufficient to kindle measles outbreaks and endanger elimination.

In 2005, an Indiana outbreak provided a clear example of the potential for anti-vaccine messages to leave communities vulnerable. This story begins with an unvaccinated seventeen-year-old girl who traveled with a church mission to an orphanage and hospital in Bucharest, Romania. Although Romania had MCV1 coverage above 95 percent among children, it recorded more than 5,000 measles cases in 2005, reflecting pockets of undervaccination among older children and in marginalized communities. In that setting, orphanages, crowded with unimmunized children, were at particular risk. Despite developing a fever, a cough, a runny nose, and red eyes prior to her flight, the Indiana teen returned home on May 14 and attended a church gathering with several hundred people the next day. Two weeks later, a six-year-old girl who had been at the event was hospitalized with measles. Contact investigations turned up thirty-four confirmed cases of measles, all stemming from the returned traveler. Most cases were in those under twenty years of age, and 94 percent were unvaccinated against measles. Indiana had school vaccine mandates in place, but most of the affected children were either homeschooled or had obtained nonmedical exemptions, allowing them to avoid the requirement. Of the two cases who had been vaccinated against measles, one was an adult who had had a single dose of a measles-containing vaccine in childhood. She worked as a phlebotomist at a hospital where two of the cases were admitted and had been exposed in the course of her job. Unfortunately, she required intensive care, including six days on a mechanical ventilator, as a result of measles complicated by pneumonia.

CDC and local health authorities held follow-up interviews and focus groups with church leaders and parents to explore the Indiana community's attitudes about immunization. Unlike the situation in the Philadelphia outbreak, the involved church had no official position on immunizations, and none of the parents reported receiving information about vaccines from the church prior to the outbreak. What was clear was that Wakefield's message—wrong but sticky and contagious—had not only made it to the United States but had been hard at work in this Indiana community long before the Romania mission and long before anyone had a visible measles rash. The most common reasons that the Indiana parents gave for vaccine refusal were media reports about potential harmful effects of vaccines, especially autism, a preference for "natural" over vaccine-induced immunity, and their own personal religious beliefs (despite the reported absence of input from church leaders). Many families with children who developed measles in the outbreak had sought information from alternative health care providers who doubted the wisdom of the recommended vaccine schedules. The Indiana outbreak accounted for more than half of the cases of measles in the United States in 2005, and control of the outbreak led to more than $160,000 in costs. It was just the beginning.

Wakefield didn't wait to see if his message would make it to Minnesota. He brought it there himself, targeting an isolated immigrant group. The largest Somali American community in the United States is in the area near the twin cities of Minneapolis and St. Paul, Minnesota. In 2004, the rate of MMR vaccination among Minnesota children was high—above 90 percent. But looking at a graph of these rates over time reveals a striking change starting around 2008. The lines representing the children of Somali and non-Somali families separate, with the vaccination rate among Somali children dropping significantly and the other line remaining stable. By 2010, the difference is stark—with only about half of Somali two-year-olds receiving an MMR vaccine. This simple picture tells a story of frightened parents, community distrust, and targeted anti-vaccine misinformation that culminated in the first of two measles outbreaks concentrated in that community.

In 2008, Somali families raised concerns that too many children from their community were receiving diagnoses of autism and were being placed in special education programs in the public schools. Increased screening and availability of services for autism led to rising numbers of diagnoses in many areas. City officials met with parent groups and pledged to study the problem in conjunction with the school system and epidemiologists from the University of Minnesota. As the researchers geared up to gather data and approach the problem scientifically, anti-vaccine activists moved in. J. B. Handley, a businessman turned anti-vaccine zealot, communicated with concerned parents over the course of several months and subsequently posted an open letter on the Age of Autism website, addressing the "Courageous Somali Parents." His message in the letter was simple—public health authorities were not to be trusted. Handley, who is not a doctor, provided clear advice about medical matters to the members of the community, suggesting that pregnant women should avoid vaccines and dental work and that childhood vaccines should be spread out by at least three months. As for the MMR, his opinion was that parents should get their children separate measles, mumps, and rubella vaccines. None of these recommendations has any basis in science. In addition, the individual measles, mumps, and rubella components are not available in the United States, and there is no medical reason to separate them.

Handley is a man who takes his anti-vaccine work seriously. In 2005, he founded a group called Generation Rescue, which, with the help of celebrities such as Jenny McCarthy and Jim Carrey, has fought to promote Wakefield's debunked theories and to sow doubt about vaccine safety among the public. Generation Rescue cosponsored a "Green Our Vaccines" rally in Washington, DC, in 2008, where Robert F. Kennedy Jr. delivered a keynote address. Their motto was that the vaccine schedule was "too many too soon," and McCarthy claimed that "We're not an anti-vaccine movement . . . we're pro–safe vaccine." Wakefield himself visited the Minnesota Somali community at least three times over several months starting in December 2010, reinforcing parents' fears about MMR. By the time the last of those visits happened in late March 2011, a measles outbreak had already begun.

As was true with Indiana in 2005, the 2011 Minnesota measles outbreak started with an unvaccinated child returning from international travel. The slow sowing of anti-vaccine attitudes was the kindling; the returning traveler was the spark. Exposed community members became ill, and over a two-month period there were twenty-one cases, most of whom were unvaccinated. Two-thirds required hospitalization, and the health department identified more than three thousand exposed individuals. Contact tracing—a vital component of outbreak control—was particularly challenging, given that some of the earliest transmissions took place in a drop-off day care center without a stable roster of children (or a system to collect immunization records), and others occurred within two shelters for people experiencing homelessness.

The first diagnosed case, who was not the child who traveled, was a nine-month-old residing in one of the shelters. Six cases were children from the Somali community who were unvaccinated because of parental concerns about autism. For his part, Wakefield disavowed any role in the outbreak, saying, "I don't feel responsible at all."

That outbreak was a warning sign, and measles was not finished with the Somali community. MMR vaccination rates continued to plummet, reaching 42 percent among two-year-olds in 2016—a rate much lower than the 90 to 95 percent required for effective community protection. Seventy-five total cases were diagnosed over five months. More than 80 percent were members of the Somali community. Nearly all of those affected were unvaccinated, and transmissions concentrated in child care centers and homes. Within single households, the measles attack rate among unvaccinated contacts of a diagnosed case was above 90 percent. Public health officials identified and investigated more than eight thousand contacts, and the total cost to control the outbreak topped $2 million. As vaccination rates dropped, outbreaks involved more children, went on longer, and became more expensive to control—even when they happened in the same place. The Minnesota Somali community continues to feel the effects of the anti-vaccine movement's targeting. In 2018, the MMR vaccination rate among two-year-olds in the community fell below 35 percent, and measles outbreaks continue to strain the local public health system and endanger the community's children.

In the United States, the period from 2014 to 2015 also brought a spike in measles activity, and even more so in measles publicity. An increase in measles cases in 2014 (667 reported for the year—about three and a half times the number in 2013) was mostly driven by an outbreak in an Amish community in Ohio, but the story that captivated the news cycles started in December 2014. Three days after Christmas, an unvaccinated eleven-year-old child developed a rash. That child ended up requiring hospitalization, and on January 5, 2015, the California

Department of Health began to investigate. The child had not traveled outside the country. However, he had been at Disneyland in mid-December. Before the day was over, reports of six additional suspected cases arrived—all six children had visited Disneyland around the same time. Over the next several weeks, the pace of reports accelerated, and a media uproar ensued. At least five park employees developed measles. Orange County, where the outbreak was centered, barred unvaccinated children from school.

By February 8, a CDC publication confirmed 125 outbreak-linked cases, with new reports still coming in. Many of those who were ill had visited the park, but not all of them. Subsequent waves of infection had begun among the contacts of those who had been at Disney. Media coverage of the outbreak was local, national, and international. Because many of the affected children, including the first identified case, were unvaccinated or undervaccinated, stories focused on anti-vaccine sentiments in the community. Dr. James Cherry, a UCLA pediatric infectious diseases specialist, said that the outbreak was "100 percent connected" to the anti-vaccine movement. The situation did not come as a complete surprise, as California was already well known to have a vaccine hesitancy problem.

At that time, parents of kindergarten-age children in California had three ways of circumventing vaccines that were mandated for school entry. The first was through a medical exemption, which is allowed in all states. This kind of exemption covers the extremely rare cases in which a child is highly allergic to a vaccine or has a medical condition such as an immunodeficiency that would make certain kinds of vaccines dangerous to receive. Essentially no children have medical conditions that preclude all vaccination. The other two kinds, grouped under the category of "nonmedical exemptions," allowed families to opt out on the basis of either religious or personal/philosophical beliefs. CDC data for the 2014–2015 school year indicated that 2.7 percent of California kindergartners had exemptions to mandated vaccines, with more than nine out of ten of those falling into the nonmedical exemption category. Even before the Disneyland outbreak, some investigators raised concerns that California's nonmedical exemptions had been increasing year over year—a troubling trend. Worse, it became clear that exemption rates weren't increasing evenly across the state—the nonmedical exemptions were geographically clustered. In Orange County, home of Disneyland, 8.6 percent of children had exemptions, triple the rate for the state as a whole. Most California schools had low exemption rates, but in some classes, more than a third of children were unvaccinated. These exemption hot spots were mainly private schools in wealthy, predominantly white areas. Schools that promoted alternative learning philosophies attracted vaccine-hesitant families. For example, more than 40 percent of children in kindergarten at the Waldorf School of Orange County were unvaccinated, most by parental choice. In 2011, another Waldorf school—this time in Los Altos, one of the wealthiest towns in the United States—had a kindergarten vaccination rate of 23 percent. The Waldorf schools, which focus on experiential learning and shun memorization, were founded on the ideas of anthroposophy—the belief that "humanity has the wisdom to transform itself and the world, through one's own spiritual development." These schools have become a "bastion of anti-vaccine fervor"—appealing to vaccine-hesitant parents and cultivating those attitudes within their communities.

As cases accumulated in California, *The New York Times* posted an interactive map with vaccination and personal belief exemption rates for every kindergarten in the state. The clustering of exemptions in wealthy areas around San Francisco, Los Angeles, Santa Cruz, and other cities was clear. Anti-vaccine leaders remained unswayed by the ongoing outbreak. Barbara Loe Fisher, president of the National Vaccine Information Center (NVIC)—a rebranded form of Dissatisfied Parents Together—downplayed the importance of the outbreak in the press, saying in late January that "fifty-seven cases of measles coming out of Disneyland in a country with a population of 317 million people is not a lot of cases." It is unclear whether her failure to understand the importance of local communities in measles outbreaks represented ignorance or willful omission.

Fittingly, the Disneyland outbreak also had a significant online presence. Measles-related social media activity spiked. Analysis of data from both Facebook and Twitter indicated that most of that increase was made up of pro-vaccine posts—a significant contrast to what the next few years would bring. The tenor of traditional media reporting also favored vaccination. An editorial in the *Los Angeles Times* suggested that the outbreak "should worry and enrage the public" and lambasted the anti-vaccine movement's "ignorant and self-absorbed rejection of science." Some speculated that the backlash could help bring about a "Disneyland effect," persuading hesitant parents and leading to increased vaccination rates.

A *New Yorker* cartoon drawn by Emily Flake and published at the height of the Disneyland outbreak shows a miserable-looking child sitting on an examination table, the red dots of his rash popping from the otherwise monochrome drawing. His parents—white, yuppie-dressed—look at him with concern. The father's arm is around the mother's shoulder, and his hand clasps hers. The doctor—Black, frustrated-appearing—points a gloved hand at the spots and says, "If you connect the measles, it spells out 'My parents are idiots.'" I'm not proud to admit this, as I think that mocking vaccine-hesitant parents is both mean-spirited and unlikely to do much good, but I guffawed.

By mid-April, the outbreak was over. The damage totaled 147 confirmed cases from seven additional states, Mexico, and Canada. Ominously, a traveler returning from Disneyland to Quebec seeded a separate two-month-long outbreak in a non-immunizing religious community there. Despite officially maintaining its measles elimination status, the United States had become an international exporter of contagion.

In response to the cases and the growing problem of clustered exemptions, California state senator Richard Pan—a pediatrician—cosponsored Senate Bill 277, proposing elimination of nonmedical exemptions to school-mandated vaccines. Pan had a history of advocacy in this arena, having successfully spearheaded a 2012 bill in the assembly that required that parents seeking nonmedical exemptions meet with a health care provider to discuss the risks and benefits of vaccination.

Predictably, SB 277 met with a firestorm of resistance. Parents threatened to pull their children out of school to avoid mandates. Chiropractors and other alternative health groups lobbied legislators in an attempt to derail the bill. And of course anti-vaccine celebrities let their displeasure be known. Anti-vaccine crusader Robert F. Kennedy Jr. campaigned against the bill in Sacramento and called the purported effects of vaccines on children a "holocaust." He later apologized for the comparison. In 2022, he likened vaccine mandates to Nazism. And once again he apologized. Senator Pan received death threats.

Despite the outcry, SB 277 passed and was signed into law by Governor Jerry Brown on June 30, 2015. The law took effect the next year, and while it did not revoke existing nonmedical exemptions, it did eliminate the possibility of getting new ones. The results of SB 277 have been dramatic in a number of ways, some predictable and some less so. The number of nonmedical exemptions have, of course, decreased. As anticipated and intended, vaccination rates in California schools increased significantly. However, in the absence of the option for nonmedical exemptions, the rate of medical exemptions tripled in the year following SB 277. Previously, legitimate medical exemptions had been rare and had not been rising. They also were not heavily geographically clustered prior to SB 277, in contrast to nonmedical exemptions. But as the number of medical exemptions increased, similar clustering emerged. This finding suggested that families sought out physicians who would approve medical exemptions without a legitimate indication, potentially blunting the protective effect of SB 277 against outbreaks. California health officers backed up this hypothesis, reporting their concerns about the lack of centralized review of medical exemptions, unclear rules about oversight of exemptions that appeared to be unwarranted, and physicians who charged high fees for signing off on exemptions.

In 2019, the California legislature passed a follow-up bill—SB 276—despite intensive lobbying by both anti-vaccine activists (including Kennedy) and celebrities (Jessica Biel, Rob Schneider). Like SB 277, SB 276 was coauthored by Pan. Its goal was to institute oversight of the process of obtaining a medical exemption in order to prevent fake exemptions or those from pay-for-play doctors from being approved. Biel posted on Instagram, falsely implying that SB 276 would inhibit children from obtaining legitimate medical exemptions. Protesters interrupted the legislative process. One threw a menstrual cup containing blood onto the Senate floor. Outside the capitol, an anti-vaccine activist physically assaulted Pan. Despite these challenges and despite last-minute requests from Governor Gavin Newsom to soften the exemption review requirements, the bill was passed and signed into law. Under SB 276 and its companion bill, all exemptions for students in schools with less than 95 percent of students vaccinated require review by the California Department of Public Health to ensure that they include a legitimate medical reason. In addition, exemptions from individual doctors are tracked to detect potential abuse of the system.

It is still too early to know the long-term effects of SB 276 on medical exemption rates in California, but understanding whether these legislative approaches can work and endure is crucial. Data from other metropolitan areas suggest that nonmedical exemptions based on "philosophical belief" have increased substantially in the majority of states that allow them. County-level examination of those data show that clustering of both exemptions and of decreased MMR vaccine coverage puts those areas at increased risk for measles outbreaks. In New York, where nonmedical exemptions were eliminated in 2019, medical exemptions had already climbed significantly by 2020.

SB 277 and SB 276 represented legislative defeats for the antivaccine movement, but its political power continued to grow during this period. Reality television personality Donald Trump had promoted Wakefield's bogus vaccine-autism link in the past. In 2007, he told the *South Florida Sun Sentinel*, "When I was growing up, autism wasn't really a factor . . . We've giving [*sic*] these massive injections at one time, and I really think it does something to the children." For his own child, Trump reported that they had him vaccinated on a nontraditional (and therefore untested) schedule, spacing out the vaccines over a long period of time. He continued to share his theories about vaccines on Twitter. In 2012, "Massive combined inoculations to small children is the cause for big increase in autism." In 2014, it was "Autism WAY UP—I believe in vaccinations but not massive, all at once, shots. Too much for small child to handle. Govt. should stop NOW!"

Suddenly that reality star was a candidate, and then a nominee, and then a president. In August 2016, while still a candidate, Trump met with a group of anti-vaccine donors, including Wakefield. Though the meeting was not disclosed until after he had won the presidency, Trump was described as "extremely educated" on the issues and reportedly promised the group that he would watch the anti-vaccine documentary that Wakefield had directed. In an interview with *STAT*, Wakefield said that he had been pleased to find Trump "extremely interested" and "open-minded on this issue."

After his election, Trump met with Robert F. Kennedy Jr., who claimed afterwards that Trump had asked him to head a commission on vaccine safety. Wakefield attended one of the inauguration balls and was seen on a video calling for a "huge shake-up" at the CDC. The disgraced former physician had found a kindred soul and a powerful ally, a president who had come to prominence by staking out outrageous positions, raging against establishment experts, and amplifying conspiracies on social media. Just as Trump's election newly emboldened his far-right followers, his public audiences with Kennedy and Wakefield were a clear message to the anti-vaccine movement that they had a friend in the White House. This situation would become increasingly relevant as measles cases surged in the years following Trump's election and even more so as the COVID-19 pandemic approached.

10.

Banner Years

The years 2018 and 2019 were banner ones for measles, less so for people. The disease surged globally. Countries where it had never been under control continued to have significant case numbers, but equally concerning was the loss of progress in places where transmission had once been interrupted. The region of the Americas—including all North, Central, and South American countries—achieved measles elimination in 2016, but economic collapse in Venezuela led to the crumbling of the nation's public health and medical infrastructure, leaving Venezuelans with no way to track or control outbreaks. As a result, measles transmission returned to the country in 2017, and Venezuela's elimination status (and thus, that of the region as a whole) was lost the following year.

Europe also had record case numbers in 2018, driven largely by outbreaks in Ukraine. In the years leading up to this surge, vaccination rates of Ukrainian children had dropped significantly due to a combination of vaccine hesitancy (a long-standing problem that was acutely potentiated by rumors that circulated after a teenager had died of an unrelated cause in the days after receiving a vaccine), political instability in the wake of Russia's annexation of the Crimean Peninsula, and the subsequent armed conflict, which thwarted vaccination campaigns. As a result, Ukraine had near-constant measles transmission, with more than 54,000 cases in 2018 and numerous deaths. Some of those cases occurred in visitors to the country. When those individuals returned to their home nations, they seeded outbreaks elsewhere.

In the United States, measles case numbers fell in 2016 after the Ohio and Disneyland outbreaks in 2014 and 2015, but in 2018 they ticked up again, followed by an unholy explosion of cases in 2019. That surge was mostly driven by linked outbreaks in Orthodox Jewish communities in New York City and the nearby suburbs of Rockland County, New York. The size of those outbreaks was remarkable, as was their longevity. The story of measles in that time and place encompasses many themes that are familiar from other surges in post-elimination America but also highlights additional important weaknesses in local immunization systems. Most important, New York's experience with measles in 2018 and 2019 presaged many of the challenges that the city would face in the early waves of the COVID-19 pandemic.

The measles outbreaks in New York's Orthodox Jewish communities didn't arrive without warning. As had been the case for the Somali American population in Minnesota, several recognizable elements were already in place. Smaller clusters of vaccine-preventable diseases. A growing acceptance of vaccine hesitancy. Tensions between community members and the local government. Check, check, and check.

Mumps is a viral disease that, like measles, was common among children in the United States prior to the development of a vaccine in the 1960s. It also frequently caused outbreaks in military barracks, dormitories, and other situations involving close quarters. Some children with mumps have no symptoms at all, but most have inflammation of the parotid glands, which can lead to a characteristic "chipmunk cheeks" appearance. About a third of unvaccinated males who get the mumps have painful swelling of the testicles, called orchitis. Nearly half of those with orchitis develop testicular atrophy (wasting of the tissue), which can lead to infertility. Mumps also threatens fertility in females by causing inflammation of the ovaries. While the mumps vaccine isn't perfect at protecting against infection, it makes all these complications much less likely.

In 2009–2010, a large mumps outbreak—more than three thousand cases—swept through the Orthodox communities. An eleven-year-old boy returned from a trip to the United Kingdom, which had an ongoing mumps outbreak, and developed symptoms of the disease after arriving at a summer camp in upstate New York that he attended with hundreds of other Orthodox boys. Twenty-five additional cases of mumps occurred at the camp, and at the end of the session, the campers returned to their communities in Brooklyn, Rockland County, and elsewhere in the area. Spread from those children to others in their communities promulgated the outbreak for nearly a year. Well over a hundred boys had orchitis (putting their future fertility at risk), six had mumps meningitis, and three developed deafness as a result of the infection. Religious schools (yeshivas), where intense face-to-face discussion and debates were both an integral part of the pedagogy and a likely means of transmission, were hotbeds of contagion.

One interesting facet of this particular mumps outbreak is that most of the affected individuals had received two doses of MMR vaccine. Immunity against mumps after MMR vaccination is less durable than for either measles or rubella, meaning that adolescents and young adults are at greater risk of getting the infection because it has been longer since they received their childhood MMR vaccines. As a result, outbreaks of mumps commonly occur in highly vaccinated but close-knit groups. While this kind of information is often twisted by anti-vaccine groups to attempt to convince parents that the vaccine doesn't work as it is supposed to, it is in fact a testament to why routine vaccination is so important. No tool is perfect, so some people who received two doses of MMR vaccine did develop mumps (though they were at much lower risk of complications than unvaccinated children). However, widespread vaccination, particularly among younger children, can help prevent these outbreaks from ever starting.

That aspect of the 2009–2010 outbreak illustrates that the New York Orthodox community had not been particularly vaccine hesitant at the time that the involved children—mainly adolescent boys—had received their routine childhood vaccines. However, focus groups organized by city public health personnel after the outbreak revealed that many parents in the Orthodox community had since developed concerns about vaccines, including the false but sticky link between MMR and autism and the idea that the vaccine schedule might be "too many too soon"—the same old Wakefield and McCarthy tropes. Although many of the kids involved in this particular outbreak had been immunized, anti-vaccine talking points had certainly arrived in the Orthodox communities of Brooklyn.

The spread of mumps also showcased the interconnectedness of the Brooklyn and Rockland County Orthodox groups, in this case through the summer camp, but their relative isolation from surrounding populations, as 97 percent of mumps cases occurred in members of the Orthodox community. The map of individual outbreak-associated cases has clear hot spots in the Brooklyn neighborhoods of Williamsburg, Crown Heights, and Borough Park—all areas with high concentrations of Orthodox Jewish families—and only rare, scattered cases outside of those neighborhoods.

In 2013, the setting was similar, but some of the particulars were quite different. That March, a seventeen-year-old boy, unimmunized because of parental refusal, returned to Borough Park, Brooklyn, after a trip to London. He developed symptoms of measles and was diagnosed about a week after his return. However, by then he had exposed enough people to set off an outbreak that grew to nearly sixty cases. Some community characteristics seen in the mumps outbreak of several years earlier were still at play—cases were confined to Borough Park and Williamsburg. All the cases occurred in members of the Orthodox community, but this time, all of the involved individuals were unvaccinated. Twelve of those unvaccinated cases were infants under the age of one, when the first MMR vaccine is given routinely, but all others were unvaccinated by choice—some due to complete refusal of all vaccines by their family, the others due to a desire to delay vaccination. This was a major change and a clear warning sign of potential future challenges. The children involved in this outbreak had a median age of three years, considerably younger than those in the 2009–2010 mumps outbreak. Many of them came from a small number of extended families who refused vaccines. The outbreak did not spread more widely in the Orthodox community, likely because many families there were still having their children vaccinated on schedule. However, despite its limited scope, the unvaccinated status of all the cases represented a harbinger of shifting attitudes among parents in the community. Vaccine hesitancy had taken root in Brooklyn.

Mistrust of local government, particularly city public health officials, was also on the rise within New York Orthodox communities. One area of discord was around ritual circumcision. In some Orthodox Jewish sects, male circumcision is followed by the mohel (the adult male individual performing the procedure) using oral suction to remove blood from the penis of the infant. This action, called *metzitzah b'peh*, is associated with a substantially increased risk of infection for the infant and is not a routine part of the circumcision ritual for most Jewish families. Between 2000 and 2011, eleven male infants in New York City developed severe herpes infections as a result of this procedure; two of those children died from the infection. The herpes simplex virus had likely spread from the mouth of the mohel (many adults carry herpes and can shed the virus in secretions even without having any symptoms) and contaminated the open wound of the infant. Mayor Michael Bloomberg pushed for restriction of *metzitzah b'peh*. In 2012, the New York City Board of Health passed a regulation that required written informed consent from the parents before *metzitzah b'peh* could be performed. Some members of the Orthodox community reacted with outrage and threats of lawsuits. The regulation was enacted but inconsistently enforced. Mohels continued to perform the procedure, and as a result, children continued to acquire life-threatening infections. The Board of Health ruling was formally rescinded in 2015, fulfilling a campaign promise that the new mayor, Bill de Blasio, had made to advocates of *metzitzah b'peh*. Fallout from the battles over this issue, including widespread community distrust of city government (and of the health department in particular), has persisted, as has the procedure itself. I have taken care of newborns with herpes infections, including ones who have acquired it in this way, and we still encounter these cases today. It is a horrible disease with lifelong implications for the affected children. Our societal failure to protect them from this dangerous ritual is shameful.

This environment of isolation, political unease, and rising acceptance of pseudoscientific theories provided a clear opportunity for the anti-vaccine movement. Misinformation about vaccines was everywhere by the mid-2010s, and social media was particularly rich soil in which it could grow. However, many members of New York's Orthodox Jewish communities are prohibited from using smartphones, computers, or the internet. As a result, they were not generally finding health misinformation on Facebook or scrolling through Twitter. For that reason, anti-vaccine propaganda had to come through different channels to find its targets. If they were to be successful, those messages would have to take advantage of the ways that Orthodox parents, generally mothers, got advice about parenting.

One way that anti-vaccine groups targeted Orthodox mothers in Brooklyn was by telephone. A group of mothers established a hotline called Akeres Habayis (Woman of the House) as a forum for sharing parenting information. In an article in *Wired*, Amanda Schaffer interviewed a woman named Chany (identified in other reporting as Chany Silber), who describes her disillusionment with traditional medicine and her descent into anti-vaccine conspiracy theories after one of her children developed medical problems. Silber hosted conference calls on Akeres Habayis and invited outspoken anti-vaccine advocates to speak to the parents. One of these was Mayer Eisenstein, a critic of vaccines and promoter of unproven and dangerous hormonal treatments for autism. Hundreds of women joined Chany's calls, making them fertile ground for the anti-vaccine movement to plant its seeds.

Silber also helped start another organization—one that promoted a similar message using different tactics. The group called themselves PEACH (Parents Educating and Advocating for Children's Health), and they assembled a glossy booklet called *The Vaccine Safety Handbook: An Informed Parent's Guide*. The booklet is impressive in both its scope and its unapologetic emotional manipulation. The cover image is a drawing of a young girl sitting on an examining table in a vast, sterile-appearing doctor's office. Medical equipment is visible, but the girl is alone, without even a parent. She appears frightened—eyes wide, eyebrows raised—and her feet dangle off the edge of the table. The content is the expected rehashing of anti-vaccine tropes, but there are helpful cartoons (including a dismissive doctor who resorts to yelling when questioned), charts, and questions about autism and sudden infant death syndrome (which the group also falsely links to vaccination). Colorful icons catch the reader's eye. The handbook gives the appearance of being both research-based, with scores of references to pages on anti-vaccine websites, and community-focused, with some Hebrew writing and a separate section for halachic (i.e., having to do with Jewish law) points of interest. The group's companion publication, *All Your Vaccine Questions Answered*, has a similar design, recycling images and text from the handbook, but with more of a "frequently asked questions" format.

Other aspects of the PEACH handbook speak to its integration with anti-vaccine groups outside the Orthodox community. The list of helpful websites toward the end of the text includes those with content from prominent anti-vaccine voices such as Dr. Sherri Tenpenny, whom Chany cites as an early influence, and the NVIC. The "PEACH hotline" number that is listed is the same as Akeres Habayis. Finally, the publication's masthead lists Barbara Loe Fisher, of Dissatisfied Parents Together/NVIC fame, as a "contributing researcher," a direct connection from the early parent groups of the 1980s to the Orthodox anti-vaccine movement of the twenty-first century.

In the mid-2010s, PEACH's handbook began appearing unbidden on doorsteps, in grocery delivery bags, and through the mail. Tens of thousands of copies were distributed, and there were mass mailings to Jewish communities in other cities. And it worked. The slow fall in vaccination rates accelerated. In New York City, rates of religious exemptions from required vaccinations among students in Orthodox Jewish schools nearly quadrupled—from 0.7 percent in the 2012–2013 school year to 2.7 percent in 2018–2019, ten times the rate in New York City public school students. In Orthodox-rich Williamsburg, the percentage of eligible children under age five who had received a measles vaccine was under 80 percent, much too low to prevent or stop an outbreak. Outside the city, in the Orthodox communities of Rockland County, mothers had also received PEACH's messages—nearly one in four schoolchildren was unvaccinated.

Under such conditions, it was just a matter of time before measles returned to these communities. In the autumn of 2018, it did. The chain of events leading up to the New York outbreaks starts in Ukraine, where measles raged under conditions of war and vaccine hesitancy. On the occasion of Rosh Hashanah, the Jewish new year, members of Hasidic Jewish sects set out on an annual pilgrimage to the town of Uman in central Ukraine. The purpose of the journey is to visit the grave of Rabbi Nachman of Breslov, a spiritual leader credited with reinvigorating the Hasidic movement in the late 1700s and early 1800s. Each year, tens of thousands of people participate in the pilgrimage. In 2018, Rosh Hashanah lasted from September 9 to 11. There had been measles cases in Israel earlier in the year, mainly in the northern part of the country, with some in the larger cities. However, after the return of large numbers of travelers from Ukraine later in September, measles took off in larger Israeli cities, including both Tel Aviv and Jerusalem. Both cities had measles vaccine coverage of more than 90 percent among children, but the rates in Orthodox enclaves were much lower. By late September, there was substantial transmission among the unvaccinated in both places.

Travelers brought measles from Israel to the United States, where cases were identified nearly simultaneously in Brooklyn and Rockland County. The first child diagnosed in Brooklyn had returned from Israel on September 21, and their first day of rash was September 30. In Rockland, the first diagnosed case was an unvaccinated teenager who had arrived on September 28. He had attended a crowded synagogue five times over four days prior to his diagnosis on October 1, potentially exposing large numbers of people. Over the next several weeks, at least six additional cases emerged among unvaccinated travelers from Israel shortly after their return. The implication of that timing is that many of the travelers were exposed while in Israel, only becoming ill once they had returned to the United States. Thus, unlike most of the other post-elimination outbreaks, the 2018 outbreak began with multiple independent introductions of measles from a country with ongoing spread, rather than a single discrete importation.

Over the following weeks, cases accumulated in both Rockland and Brooklyn. Local health departments printed up flyers, held vaccination drives, went door to door. By mid-January 2019, there were 116 confirmed cases in Rockland. Fifty-eight and counting in Williamsburg and Borough Park. Both areas instituted school restrictions, ordering exclusion of children who were unvaccinated against measles, regardless of the reason. Some parents consented to vaccination at this point, but others fought the orders. Parents of more than forty students at Rockland's Green Meadow Waldorf School, where less than a third of students were up-to-date on their immunizations, sued, asking a judge to allow their unvaccinated children to return to classes. They were unsuccessful, as challenges to vaccine mandates generally have been since 1905, when the Supreme Court decided *Jacobson v. Massachusetts*. A Williamsburg school, Yeshiva Kehilath Yakov, refused to exclude unvaccinated students. In January, one contagious student there infected more than twenty others.

By February 2019, the Brooklyn outbreak showed no signs of abating, and it felt like measles was everywhere, not just in New York. Health officials in Clark County, Washington, an area with high rates of vaccine refusal, declared a state of emergency due to accelerating cases among unvaccinated children. (The index case for that outbreak was an unvaccinated ten-year-old who had traveled to Ukraine.) On the heels of dropping MMR vaccination rates due to parental fears, the Philippines reported fifty-five measles deaths among children in the first month of the year. The World Health Organization listed vaccine hesitancy among its top ten threats to global health in 2019.

And then, similar to the reaction to the Disneyland outbreak, it seemed that people might finally have had enough of the anti-vaccine crowd. Articles and editorials specifically calling out both the lies and the destructive potential of the anti-vaccine movement appeared more frequently. The story of Ethan Lindenberger, a teenager who got vaccinated in defiance of his mother's staunchly anti-vaccine beliefs, captured national attention, and he was invited to testify in front of a Senate committee. Technology companies, including Pinterest, YouTube, and Facebook, announced plans to begin to combat misinformation about vaccines online. On March 6, 2019, three Trump administration officials published a joint opinion piece titled "This Is the Truth About Vaccines" denouncing the purported link between vaccines and autism. No mention was made in the article about the president's feelings on the matter.

Despite this apparent shift in rhetoric and ongoing health department efforts to convince hesitant families to vaccinate their children, cases continued to spike. Rockland's measles vaccination rate remained below 80 percent. On March 26, county executive Ed Day declared a state of emergency, barring minors who had not been vaccinated against measles from public places. Day was clear that the order was intended as an "attention grabber" and that his plan was not to make arrests. Legal challenges and vitriol followed. There were also concerns from people who favored school mandates and vaccination in general but wondered whether the order was pragmatic and whether distrust in the community might deepen as a result.

In early April, New York City followed Rockland's lead. At that point, the city had 285 confirmed measles cases, and the outbreak had accelerated. Mayor de Blasio declared a public health emergency, requiring individuals who lived, worked, or attended school in high-risk areas of Brooklyn to be vaccinated against measles or face fines. De Blasio warned that under the state of emergency, yeshivas that continued to flout public health orders and allow unvaccinated students to attend would be shut down. Opponents of the measure circulated affidavits and planned for lawsuits, and tensions in the city rose. PEACH held a four-hour conference call, using multiple speakers with important-sounding titles to push out anti-vaccine messages to families. Orthodox families received robocalls asking them to fight against potential restrictions on unvaccinated individuals in synagogues and yeshivas. On April 15, the mayor made good on his promise. The New York City Department of Health and Mental Hygiene shut down a preschool program at the United Talmudical Academy in Brooklyn, which had refused to allow audits of vaccination and attendance records. Additional school closings followed.

The mood in New York City was one of frustration on all sides. Parents of those who were too young to be vaccinated reported keeping their children out of day care and avoiding events where unvaccinated children might be present. A non-immune pregnant woman walked an eight-mile round trip to work to avoid possible exposures on a Brooklyn subway. (Live vaccines, including the measles vaccine, are not generally recommended during pregnancy.) Parental disagreements over vaccination became issues in divorce proceedings. Incidents of anti-Semitism connected with measles rose in New York City, and Orthodox leaders worried about escalation as the outbreak dragged on. Moshe Friedman, a Hasidic man, called out members of his community for their "terrible mistake about vaccinations," which he attributed to a combination of deliberate misinformation and a lack of scientific literacy. (Rabbis from the community were overwhelmingly supportive of vaccination, and there is no prohibition in Judaism—or any other major religion—against vaccination.) Pediatricians, including me, expressed frustration at fighting a disease that we already knew how to prevent.

And then there were the surreal bits. Sure, measles popped up in Los Angeles, requiring nearly a thousand college students to quarantine. But even outside the United States, it seemed like measles stories were getting stranger. Germany proposed fines of 2,500 euros for parents who failed to vaccinate their children. The *Freewinds*, a cruise ship owned by the Church of Scientology, was quarantined by the nation of St. Lucia after attempting to land with an active measles case on board. Someone wrote to an advice columnist to complain about their sister's anti-vaccine beliefs and to ask whether it would be ethical to take the sister's children to get vaccinated without their mother's knowledge. (Reader, their answer was no.)

In mid-May, the anti-vaccine movement doubled down. With cases still rising nationwide, a "vaccine symposium" in Rockland drew hundreds of attendees through robocalls, flyers, and word of mouth. The lineup of speakers featured local anti-vaccine personalities such as Lawrence Palevsky (a pediatrician who eschews scientific studies of vaccine safety) and Rabbi Hillel Handler (an Orthodox leader from Brooklyn who is an outspoken critic of vaccines). Palevsky, who has no formal training in infectious diseases or vaccine science, falsely claimed that bad lots of vaccine were being given to Jewish communities and were causing measles outbreaks. Rabbi Handler invoked persecution, claiming that "the campaign against us has been successful." In an odd turn of events, he also falsely claimed that Mayor de Blasio was German and was a "very, very sneaky fellow." The real star power of the night came from Andrew Wakefield himself, who addressed the crowd via videoconference, falsely denying his involvement in scientific fraud and claiming that others targeted him because he represented a threat to pharmaceutical companies.

Despite the headlines and rising anxieties, not everything that happened in the outbreak was either a governmental order or anti-vaccine bombast. New, optimistic voices brought hope that calmer minds might eventually prevail. Blima Marcus, an oncology nurse practitioner from Memorial Sloan Kettering Cancer Center who also happened to be an Orthodox mother living in Borough Park during the outbreak, exemplifies the potential of this approach. In the autumn, as the outbreak was beginning, Marcus was asked by her cousin, who lived in an Orthodox community in New Jersey, to join a group text chat of about forty mothers, many of whom were hesitant about measles vaccination. She found that misinformation was prevalent within the group, with many mothers having read the PEACH booklet and other targeted messages. Marcus addressed their concerns respectfully, calmly, yet firmly, presenting information she had gained from reading scientific studies.

Feedback from women in the group chat was positive, but misinformation was still rampant in the Orthodox communities. Marcus joined forces with other members of the Orthodox Jewish Nurses Association to form a task force on vaccines, which eventually became the Engaging in Medical Education with Sensitivity (EMES) Initiative. In Hebrew, the word *emes* means truth or honesty, emblematic of her approach. EMES members organized small gatherings of Orthodox women in living rooms and classrooms. Their discussions were civil and fact-based; they brought scientific articles and displayed graphs of vaccination rates and autism diagnoses for the attendees to analyze together. She describes a major goal of the group as educating parents about separating quality information from misinformation. Their points were based on science, with explicit discussion of the impact of decisions on both individuals and members of the community. Progress was hard-won, measured in individual mothers.

One of the major challenges that the members of EMES faced was having to counter misinformation delivered through other sources, including the anti-vaccine conference calls and meetings. Marcus and her colleagues were particularly vexed by the PEACH booklet, which seemed to appear suddenly in communities in advance of their meetings. They decided to compile their own information into an evidence-based, heavily referenced magazine, written at an eighth-grade reading level, with strategies for critically evaluating anti-vaccine claims, information about vaccine-preventable diseases, and lists of reliable sources. The new publication was called *Parents Informed & Educated* (*PIE*), a sly reference to something one might make out of peaches. The New York City health department was impressed with the successes of EMES meetings and the favorable press response to the organization's activities. They requested that EMES produce an abridged version of *PIE* focused on the MMR vaccine (*A Slice of PIE*) and distributed nearly 100,000 copies. The more personal EMES approach seemed more effective than governmental proclamations.

As June approached, the outbreak began to wane in both Brooklyn and Rockland. Vaccination levels had risen—tens of thousands of MMR doses had been given in the affected areas. In New York City as a whole, more than 185,000 doses were distributed over the course of the outbreak. The cases stayed largely confined to unvaccinated populations in both areas. High immunization rates outside of those communities and decreasing numbers of susceptible hosts within them (due to a combination of some individuals getting vaccinated and others developing protection after getting the measles) led to the virus having fewer new people to infect. The value of R—the measure of measles's ability to spread under the current conditions—dropped, as it does under such circumstances, and the curve sloped downward.

EMES continued their work even after case rates had decreased. They organized a large educational event in Rockland County in June, with nurses staffing booths and distributing information. Workshops for physicians and other health providers helped spread their techniques for countering misinformation beyond the organization, and Marcus and her colleagues published papers in nursing journals to spread the word to their colleagues. Once the New York state legislature, following California's lead, voted to eliminate religious exemptions (there was already no allowance in New York law for a nonreligious personal belief exemption) to school mandates for vaccination, EMES held another fair in Brooklyn, answering questions about the change in the law and helping hesitant parents move toward vaccination in advance of the new school year.

There are important lessons in the success of Blima Marcus and EMES. Many Orthodox mothers who attended their living room meetings did not want to hear from health department officials. When public health agencies talk about putting out culturally guided messaging, that often means involving community or religious leaders. In the case of the New York outbreaks, a letter went out from five hundred Orthodox physicians in support of vaccination. Numerous rabbis provided pro-vaccine messages. There is evidence that Orthodox mothers may have seen public health professionals as arms of an untrustworthy local government, physicians as biased purveyors of information (potentially due to messaging from PEACH), and male rabbinic pronouncements as outside the framework that they would generally use to make health decisions. The slow, difficult-to-scale, hands-on EMES approach—complete with follow-up texts to check on a newly convinced mother and her freshly vaccinated kids—may be the best way to make progress.

The other lesson here—one that we truly needed to learn prior to the approaching pandemic—is that the who of public health messaging can matter. A lot. Medicine and public health voices have traditionally been white and male. Representation in these fields is broader than it was decades ago, but it's still nowhere near where it should be. That fact makes us worse at accomplishing our goals than we have to be. A more representative workforce would likely have been more effective at communicating with families about measles vaccines in the "hard to reach" populations in the late 1960s, might have understood the vaccine access problems faster in Chicago or New York or Los Angeles in the 1990s, and could have gotten scientifically sound information out faster in the pandemic 2020s. EMES contrasted with those past and future failings—it shouldn't have been surprising that Orthodox Jewish mothers were more interested in what Orthodox nurses had to say than in what others, myself included, thought.

The final numbers for the 2018–2019 New York outbreaks are staggering. Rockland had 312 confirmed measles cases, with a handful in surrounding counties as well. New York City had 649 cases, of which nearly three-quarters lived in Williamsburg. More than 20,000 contacts required investigation, and health department officials were stretched thin. The 2013 New York City outbreak, kept in check by a still-robust community vaccination rate, had cost the city about $400,000 (measured as health department costs and not including medical costs). The corresponding figure for the 2019 outbreak was more than $8 million.

Nearly 95 percent of New York City cases were members of the Orthodox Jewish community, and more than 85 percent of those with a known vaccination history had received no doses of measles-containing vaccines. Forty-nine people, overwhelmingly unvaccinated children, required hospitalization, twenty of those, including several children whom I helped care for directly, needed our highest level of care—a pediatric intensive care unit. Anti-vaccine proponents or measles minimizers make much of the fact that there were, thank goodness, no fatalities among the children in this outbreak. That has less to do with measles's intrinsic propensity to kill and more to do with the presence of medical care, including pediatric ICUs, for those who need it. Had this outbreak gone on much longer, there would almost certainly have been fatal cases in the city. One's luck can only hold out for so long.

By the end of 2019, the United States had 1,274 confirmed cases of measles, its highest total since 1992—the year right after the nationwide epidemics. Before elimination. Before the Clinton initiative and Vaccines for Children. Because both New York outbreaks lasted less than a year, the United States had once again managed to hold on to its official measles elimination status, though barely. The societal kind of immune amnesia had taken hold, even if only in some individual communities. We had taken vaccines for granted and were paying the price.

And although the United States had had its worst measles year in decades, the situation elsewhere in the world was much, much worse. In the island nation of Samoa, the MMR vaccination rate fell precipitously after a highly publicized 2018 case in which two children died when nurses mistakenly mixed MMR vaccines with anesthetic agents instead of sterile water. The government had temporarily paused MMR vaccine delivery during investigation of the incident, restarting it once it was clear that the problem was an isolated one and that MMR vaccination remained safe. However, anti-vaccine activists recognized an opportunity to promote confusion and lies. Robert F. Kennedy Jr. traveled to Samoa to meet with the prime minister and local activists, ghoulishly spreading unfounded anti-vaccine messages in an attempt to ensure that no tragedy would go to waste. In December 2019, as the earliest cases of what would come to be called COVID-19 were being detected in China, measles raged in Samoa. Over the course of several months, its population of about 200,000 had suffered nearly 6,000 measles cases and more than 80 deaths, most of which were in children under age five. Samoa did not even have enough child-sized coffins; more had to be imported.

In 2019, the Democratic Republic of the Congo saw a massive spike in measles activity, with more than 310,000 cases and 6,000 deaths. International attention had been focused on the country's Ebola virus outbreak that had started in 2018. However, measles—a perennial threat in the DRC—claimed far more lives. Ending the measles outbreak there required a massive campaign, involving multiple international agencies, and the eventual vaccination of more than 18 million children under age five.

Estimates of the global burden of measles in 2019 were discouraging—9.8 million cases, more than 200,000 deaths. The latter figure represented a nearly 50 percent increase since 2016. To be sure, measles deaths have dropped significantly over the past two decades as global vaccination coverage has increased. The backsliding in 2019 did not bring global numbers back to anywhere near where they were in 2000, much less in the time before measles vaccines were available, but the message that the trend carried was there. The world had rising vaccine hesitancy, competing priorities, and areas of armed conflict that threatened public health systems. For all these reasons, it was apparent that measles control, where it existed at all, was in a precarious state.

Of course, none of us knew then what new horrors lay right around the corner. Measles, as always, was a bellwether, illuminating the challenges that we faced, the areas in need of attention, the holes in our defenses. This time the message that it was sending was that we were losing ground in immunization delivery, trust, and public health. That ground was lost even before a global pandemic began.

11.

Booster Shots

In retrospect, measles in 2019 feels a bit like a practice run—a prelude for the issues that we would face during the coming pandemic. The crescendo of anti-vaccine voices raising doubt online and in pamphlets left on doorsteps. The increasingly vocal skepticism of doctors, public health authorities, and science itself. Fights pitting individual rights against community responsibilities. Minimization of the potential harm of a deadly virus. Schools—and even schoolchildren themselves—used as battlegrounds over vaccination. Underfunded global public health interventions leading to continued outbreaks of a disease that we had the tools to prevent. It was all there with measles, and it all fed directly into the issues that we faced during COVID-19 and continue to face today.

In a prescient opinion piece published in June of 2019—amid large measles outbreaks and months before the first COVID-19 cases—cryptographer Bruce Schneier warned that the next pandemic, whatever its cause might be, would require "fighting on two fronts." There would be the war against the pathogen, of course, with the need to develop vaccines, drugs, and containment strategies. But all of that would be compounded by the war against the inevitable fake news that would follow. He envisioned a situation in which wrong information—either the kind that is well-meaning but incorrect (misinformation) or the lies that are malicious at their core (disinformation)—would hinder pandemic responses, drive people to seek out unproven or dangerous cures, and potentially cause essential services to falter.

Essentially all of this has come to pass. Measles, as it always has, showed us our vulnerabilities, and we did little to fix them.

The early months of the COVID-19 pandemic were a time of unimaginable fear and confusion. In New York City, where I live and work, regular activities ceased. Lockdown orders were put in place, and intensive care units filled. Each morning, on my way to work, I walked past refrigerated trucks serving as overflow valves for our hospital morgues. Personnel were shuffled so that anyone who could do so was pulled to care for the sick; others stayed home. I wore the same clothes to work every day, pulled them off immediately when I got home, hung up my mask, started the laundry, got in the shower. The next day was the same. And the one after that. All of us struggled with the sometimes competing needs to care for our patients, protect ourselves, worry about our families, and figure out how to move forward. Doctors learned about this new disease on the fly. Many treatments that seemed promising early on ended up being useless when rigorously evaluated over time. The same was true for public health agencies. Some of the early advice from the CDC or local health departments ended up, in retrospect, being incomplete or incorrect. I bring up this piece of the story simply to point out that everyone was confused at the beginning. This wasn't influenza or measles or even the kinds of coronavirus that we knew about—the ones that caused colds, or even the first SARS virus, which had never spread like its successor could. New virus, new rules. Messages and actions changed over time as we felt our way through.

In that setting, it is not surprising that confidence in medical and public health institutions was shaken. At the Munich Security Conference in mid-February of 2020, Tedros Adhanom Ghebreyesus, the director general of the WHO, warned that the world was "not just fighting a pandemic; we're fighting an infodemic." COVID-19 misinformation became rampant online. Some was of the well-meaning but wrong kind, but quite a bit was the other kind. Speculation about the origins of the virus, including theories about its intentional or accidental release from a laboratory in China, stoked xenophobic rage. President Trump called it "China virus." He called it "Kung Flu." Theories that lockdowns and mask mandates were early steps in governmental control of populations led to protests and defiance of public health guidance. Early releases of reports of shoddy scientific investigations led to hucksters (and even a president) calling for widespread use of drugs like hydroxychloroquine or azithromycin—agents that did not work and had the potential to cause harm. Years later, they would still be hawked as "cures" that mainstream medicine had for some reason suppressed.

Even before there were COVID-19 vaccines available, anti-vaccine activists saw the unique potential in this time of confusion and fear. Journalist Tara Haelle quoted anti-vaccine advocate Joshua Coleman speaking about this opportunity at a rally in March 2020, urging attendees to capitalize on this unique moment to disseminate anti-vaccine information. Members of that movement had already picked up on the power of using populist arguments that focused on individual rights and parental autonomy as means of fighting against vaccine mandates. Thus it was no surprise when the Freedom Angels, an anti-vaccine collective forged in the battles over SB 277 in California, joined forces with far-right extremists and members of armed militias at protests directed against pandemic-related restrictions in multiple states. Anti-mask and anti-lockdown demonstrations relied heavily on arguments of the sanctity of individual choice and represented a logical link between anti-vaccine rhetoric and distrust of COVID-19 mitigation measures.

Once COVID-19 vaccines became available, there was a palpable outpouring of hope. I cried when I got my first dose, overwhelmed by joy, relief, and exhaustion. I cried again when my wife got hers, and when my daughter got hers. Like many of us, I thought that the nightmare might be ending and that science—vaccine science—had saved us. I remember thinking that, like the polio vaccines before them, the COVID-19 vaccines might serve as an unmistakable demonstration of the power of science to do good. It also occurred to me that this triumph might be a mortal wound to the anti-vaccine movement. I was naive.

Part of the problem was that we—physicians, scientists, public health experts—messaged imperfectly, sometimes poorly. In the early months, COVID-19 vaccine doses were scarce and were prioritized differently in different states, leading to confusion and frustration. Vaccine availability also varied greatly from country to country, and not in a way that corresponded to the toll of the disease. We still have a lot to learn about equity. Initial studies suggested that two doses of an mRNA vaccine provided impressive protection. Warnings that we didn't know exactly how long that protection would last were drowned out by hopeful messages. As a result, when recommendations for booster doses came, they were viewed with suspicion—as if a promise had been broken—and uptake was low. Arguments erupted about whether children should be vaccinated against a virus that caused severe disease much less often in the young than in the old. My colleague Dr. Perri Klass and I wrote a piece in early 2021, using measles as a point of comparison for the debates over COVID-19 vaccination for children. We argued that vaccinating children was both an ethical obligation (to protect them from disease) and a practical necessity (because population-level control of the pandemic would require it). Our point of view was not universally shared. Dr. Michelle Fiscus, a pediatrician who was serving as the top vaccine official in Tennessee, was targeted by anti-vaccine groups and fired for her vocal support for vaccinating children. Even in the face of rising pediatric hospitalizations and the appearance of an unusual post-COVID-19 syndrome called MIS-C (multisystem inflammatory syndrome in children), the narrative of COVID-19 as a disease that exclusively affected adults held. This early opposition to pediatric vaccination was itself sticky, and even years later, acceptance of COVID-19 vaccines for children remains far lower than it should be.

Anti-vaccine personalities, particularly Robert F. Kennedy Jr., had already been trafficking in COVID-19 conspiracy theories, but when the vaccines became widely available—and particularly when calls for vaccine mandates grew—they became even more outspoken. In 2021, Kennedy's organization, Children's Health Defense, released a film called *Medical Racism: The New Apartheid*, which targeted anti-vaccine messages and COVID-19 falsehoods specifically to Black Americans, a group that suffered disproportionately during the pandemic, the fallout of decades of systemic racism. Over the course of the pandemic, Kennedy has attended and spoken at multiple anti-vaccine and anti-mandate rallies. His book *The Real Anthony Fauci: Bill Gates, Big Pharma, and the Global War on Democracy and Public Health* promotes the false thesis that Fauci, who was the head of the National Institute of Allergy and Infectious Diseases and a leader of the national pandemic response, and Gates are at the helm of a global cabal spreading risky vaccines and using COVID-19 for population control. Kennedy used the notoriety boost from his anti-Fauci posturing to jump-start his own presidential campaign.

The effect of prominent anti-vaccine voices promoting pandemic conspiracies, underplaying the potential severity of COVID-19 as "just a cold," and sowing doubt about the safety and effectiveness of vaccines to prevent it is at least twofold. One effect is that uptake of COVID-19 vaccines has been less than optimal, particularly among adults under age sixty-five and children. Because the vaccines are safe and confer substantial protection against severe disease and death, this situation translates directly into more hospitalizations and deaths due to COVID-19 than need to occur. The second effect is more insidious and is visible in Coleman's early pandemic comments about education. Anti-vaccine groups have weaponized the politicization of discussions about vaccines and public health, the confusion that the pandemic has brought, and the autonomy-based arguments about lockdowns. All of these forces are now focused on their true target: routine childhood vaccination. Alex Shephard called COVID-19 vaccine skepticism a "gateway drug" for more extreme anti-vaccine views, with parents and lawmakers primed by the COVID-19 fights questioning the wisdom of all vaccines and pushing back against all mandates. Robert F. Kennedy Jr. never lost sight of that goal, either, leveraging public doubt about COVID-19 vaccines to consolidate opposition to routine childhood vaccines, especially MMR. Kennedy wrote the foreword to Children's Health Defense's 2021 *Measles Book*, which promises to reveal "secrets" about the disease and the vaccine. Spoiler alert: There are no secrets in the book, just the same old disproven anti-vaccine tropes.

Even as the pandemic wanes, resistance to routine vaccinations persists. In late 2022, a Kaiser Family Foundation survey found that more than a third of parents oppose vaccine mandates for school attendance. In 2019, the comparable rate was under a quarter, meaning that the pandemic period has brought a significant increase in resistance to a vital public health tool. American anti-vaccine attitudes now put pets at risk as well, with a rising number of owners refusing rabies vaccination for their dogs.

Heidi Larson, the head of the Vaccine Confidence Project, conceptualizes vaccine hesitancy as a "trust problem" rather than an "information problem." She points to the destabilizing social effects of both the pandemic and the mistrust of institutions, particularly by members of disenfranchised communities, as driving forces in modern hesitancy. In this framework, the chaos of the pandemic and the constant hum of conspiracy theories together feed into a growing culture of vaccine hesitancy, spreading far beyond COVID-19. The state line through the heart of Texarkana determined who got measles and who didn't in the 1970 outbreak. Our current political lines, some physical and some ephemeral, have begun to determine who dies from contagious diseases. The downstream effects of these decisions are already apparent. Since COVID-19 vaccines have become available, registered Republican voters (a group with higher rates of vaccine hesitancy than their Democratic counterparts), have had a substantially increased risk of death from COVID-19. Political affiliation has become a medical risk factor.

The unholy alliance of anti-vaccine voices and the far right has hindered COVID-19 vaccination uptake and stoked vaccine hesitancy in general. In 2023, a judge ordered Mississippi to allow nonmedical exemptions to school vaccine requirements, overturning a 1979 state supreme court decision. Mississippi, a poor performer on national health indicators, has for many years been at or near the top of childhood vaccination rankings because of strong school mandate laws. This order, sparked by a lawsuit funded by the Texas anti-vaccination group Informed Consent Action Network, threatens to undermine decades of progress and to place Mississippi's children in wholly preventable danger.

Local protests have grown as well. In May 2023, I spoke about vaccine-preventable diseases at a health care conference in Brooklyn organized by leaders of the EMES Initiative. I was shocked to see protesters outside the conference venue, particularly the one who had a sign with my name on it, reading: "You do not belong in the Jewish community." My reaction, which I wouldn't necessarily have anticipated, was more sadness than fear.

The pandemic itself has created unprecedented challenges for immunization systems across the world. In the United States, the early waves of COVID-19 led to stay-at-home orders that inhibited distribution of routine childhood vaccines. Clinics and pediatric offices closed or would see only emergency patients. Parents delayed routine appointments to decrease the chance of exposure to COVID-19. An important measure of immunization delivery—orders from the Vaccines for Children program—dropped precipitously in March and April of 2020.

Soon afterwards, fewer than half of Michigan infants in some age groups were up to date on their vaccines. In Ohio, MMR vaccination rates among sixteen-month-olds fell by about ten percentage points from pre-pandemic levels. Similar patterns emerged in Texas. Fortunately, U.S. immunization rates largely recovered with the easing of local lockdowns, and in a survey of pediatricians, most reported being back to their pre-pandemic levels of immunization delivery by the end of 2020. However, a CDC analysis showed a drop in MMR coverage among U.S. kindergartners during the 2020–2021 school year to well below the required 95 percent threshold for measles protection, with several states below 90 percent. The implication of these statistics is that the effects of pandemic-related disruptions on protection against measles and other diseases in the United States may be long-lasting. Unfortunately, we are already seeing the effects of decreasing immunization rates. At the close of 2022, a measles outbreak in the city of Columbus, Ohio, reached eighty-five cases, nearly all in unvaccinated children. More than 40 percent of those affected required hospitalization. By early 2024, multiple states were once again fighting measles outbreaks in undervaccinated communities, a disheartening legacy of the COVID-19 pandemic.

The evidence of our vulnerability to vaccine-preventable infections goes beyond measles. In June 2022, an intentionally unvaccinated young adult from Rockland County was diagnosed with paralytic polio. He had not traveled internationally, meaning that his infection was acquired in the United States, something that had occurred only one other time since the late 1970s. Poliovirus is excreted in stool, and testing of wastewater from sewage systems in Rockland, its neighboring counties, and New York City revealed numerous positive samples over a several-month period, suggesting ongoing community spread. Paralytic polio is a rare outcome of poliovirus infection (most people who are infected by the virus have no symptoms at all), making this young man's case a warning sign of the virus's presence. As routine vaccination continues to be threatened, we may see more cases of paralytic polio—a dangerous return to the pre-vaccine era. Oh, and Hib? That disease that I mentioned in the introduction that my mother had seen repeatedly during her pediatric training, and I never had? In late 2022, decades into my career, I took care of my first patient with Hib meningitis, an eminently vaccine-preventable disease. It is hard to convey the sadness and frustration that went along with explaining that particular diagnosis to that unvaccinated child's parents.

Globally, COVID-19 caused deficits in immunization coverage from which it will be even harder to recover. Other high-income countries suffered disruptions to routine immunization systems similar to those in the U.S. MMR coverage in the United Kingdom fell below 90 percent for two-year-olds over the course of the pandemic. The situation in low- and middle-income countries, particularly those in the Global South, was far worse. COVID-19 disrupted both routine and supplemental immunization campaigns, and recovery, if it came, was slow and incomplete. Redirection of resources to control the spread of COVID-19 and treat the sick competed with immunization delivery for limited personnel and resources, with the result that global inequity in vaccination levels worsened over the course of 2020. WHO and UNICEF estimate that more than 23 million children missed out on basic immunizations through routine immunization systems; approximately 17 million of those received no vaccines at all over the course of the year. Global MCV1 coverage dropped from 86 percent in 2019 to 83 percent in 2020 and to 81 percent in 2021, representing 25 million children who were left without protection from measles.

Vaccination rates are about more than just protecting children against a handful of specific diseases. They are a fundamental piece of child health, a clear indicator of whether we are providing children with the most basic building blocks of a healthy future. When we are distracted by a pandemic, a war, or some other emergency, wealthy areas recover faster. We lost ground on vaccination rates in the United

States in the early months of the COVID-19 pandemic but have regained most (though not all) of it since. Areas where sporadic mass immunization campaigns can mean the difference between no protection and some protection for children, as always, bear the brunt of pandemic-related disruptions.

Of course, it's not only about vaccines. Health services beyond vaccination were disrupted. Millions of children missed out on vitamin A supplementation during 2020 as a result of the pandemic. Children who are vitamin A deficient are at substantially higher risk of measles complications, meaning that lacking both supplementation and measles vaccine is a particularly high-risk situation.

The data also illuminate a pandemic-driven collapse in global measles surveillance. Undercounting is always a problem when trying to understand how infectious diseases move through populations. Often, what we see is only "the eyes of the hippopotamus"—a tiny glimpse into a much larger disease burden. But if your counters and your lab personnel are distracted by a different pandemic or are sick themselves, even that imperfect information may simply never be tallied. Strains on local public health systems have meant that cases have gone unreported, specimens have not been sent to regional laboratories for confirmation and sequencing, and less than a third of countries achieved the sensitivity threshold for reporting that helps confirm that surveillance systems are functioning well. For those reasons, global estimates of measles cases in the time of the COVID-19 pandemic are certainly underestimated—we just don't know by how much.

Unsurprisingly, under these conditions of falling immunization rates and long-lasting disruptions of basic health services, measles outbreaks have erupted in some of the countries with the lowest MCV1 rates in 2021 and 2022. WHO called the COVID-19 pandemic a "perfect storm of conditions" for measles, with a nearly 80 percent rise in infections in the early months of 2022. Somalia, Nigeria, Yemen, Afghanistan, and Ethiopia—all with MCV1 coverage under 70 percent—each had thousands of cases in the mid-pandemic period. The synergistic impacts of conflict, crowding, and malnutrition are felt in many of those countries, amplifying the risk to unimmunized children. In 2022, more than seven hundred children died of measles in Zimbabwe due to a massive outbreak driven by vaccine hesitancy. Some Apostolic churches in Zimbabwe disavow all medical care and encouraged their members to refuse vaccines, even in the setting of a widespread outbreak. Many of the children who died from measles came from families who belong to one of those church groups. One family lost seven children to measles over a two-week period. The massive loss has caused some families to seek vaccination for their children in opposition to the church's teaching and has even led some Apostolic churches to reconsider their stances on vaccines.

Where can we go from here? The disruptions to public health systems brought by the COVID-19 pandemic will not go on forever, but we were losing ground on measles control before we even heard of COVID-19. Are we doomed to cycles of measles surges, followed by vaccination campaigns to stabilize the situation, followed by decreased funding and interest until the next surge comes along? Are we content to let the growth of anti-vaccine propaganda continue until the funding and the legal tools that are the lifeblood of childhood immunization systems disappear? To risk losing control of polio again? Even at the high point of measles control, when it was eliminated from the region of the Americas, it remained a constant threat, and our victory there was short-lived. Measles represents a constant drain on our resources and our attention, and as long as there is poverty or crowding or malnutrition anywhere, it will remain a major killer of children. If polio reemerges on a global scale, we will be forced to refight that battle as well.

Contrary to how the previous paragraphs may sound, I remain an optimist. As a pediatrician, you kind of have to be one. I'm a careful optimist, though, the kind who sees broken bones when he looks at a backyard trampoline. The kind that checks furniture in other people's homes to make sure it has been secured to the wall. (I'm fun at parties.) I'm a fan of window guards and vaccines and car seats. A turner-in-er of pot handles on stoves. An optimist who believes that with attention and care and the tools we have, we can make the world safer for children tomorrow than it is today. As a result of this outlook, I still see measles as a solvable problem. Even better, I think that we can scale the tools that we use to solve this issue to help us address other threats.

A fundamental issue in how we think about measles is memory. In the same way that the measles virus kills the cells that are the keepers of immune memories, our temporary successes and competing priorities distract us from the toll that it continues to exact every day of every year. When we forget, measles thrives—both within the body of an individual and in a society making decisions about whether to prioritize vaccination. Both kinds of amnesia leave us vulnerable to a host of conditions beyond measles. Measles is a master at infiltrating and revealing the cracks in our human systems. When we forget to use the tools that we have, when we allow antiscientific voices to influence policy, when we ignore the places that it is difficult to deliver care—there it is. For all these reasons, you can think of measles cases as a probe for whether we are paying attention. As outbreaks pop up, we should ask ourselves what we could have done better and what other warning signs we are ignoring. While measles is often first, it is a harbinger of problems to come.

Vaccines work by teaching the immune system—providing the information that it needs to defend against a future threat. Often those lessons need multiple doses over time to work effectively. Long-term memory sometimes requires booster shots—reminders for our immune system and our society. For us to deal effectively with measles, with polio, with COVID-19, and with the next pandemic that we can't even see yet, the most effective way to prepare is to teach and then boost.

Here are some suggestions:

We should try to be a bit more like Peter Panum, who escaped the bubble of traditional medical thinking. He traveled to remote islands, carefully collected data, and returned with a new perspective on a disease that others had viewed as pedestrian, a child's illness. Physicians, public health professionals, and nonexperts should recognize that changing situations and new data may require a comparable shift in thinking. There was considerable criticism of public health voices during the COVID-19 pandemic, some of it justified, because advice and messaging seemed to change over the course of hours. We need to own that uncertainty and exchange our desire to be reassuring and confident for a quicker willingness to convey our humility. Conversely, people on the receiving end of that information should understand that science and medicine might not have definitive and immediate answers to every question, particularly in new and confusing situations. Communication—honest, consistent, clear, and repeated—is key, as is realizing that that communication does not always need to, and sometimes should not, come from white-coated experts. Blima Marcus and EMES reminded us of the value of caring community engagement in fighting anti-vaccine misinformation. Similar strategies, some effective and some not, have been tried in an attempt to counter antiscientific voices during the COVID-19 pandemic.

Panum also differed from most of his contemporaries in that he intuitively focused on the social determinants of health, though he didn't use that terminology. He saw that the physical isolation of the Faroese, their nutritional status, and their colonial relationship with Denmark all played into the severity of measles once it arrived on their islands. Worse COVID-19 infection rates and outcomes among Black Americans compared to their white counterparts remind us of the systemic racism that is baked into our cities, our policies, and our health care system. In that way, a pandemic can be as much a symptom as a disease. The same is true of measles as it sweeps through Somalia or Yemen or Zimbabwe, polio as it persists in Afghanistan. The disease is important, but so is what its appearance tells us. Missing or ignoring the clues to our failings is inexcusable—they need to be taught, reiterated, and remembered.

Reminding people about scientific history and teaching scientific literacy are additional ways to consolidate memories of lessons that we have already learned. One of my favorite scientific papers is called "What We Don't See." In reviewing the successes of the past two hundred years of pediatrics, Dr. Margaret Hostetter provides a reminder of the incredible creativity and effort that went into fixing or preventing or minimizing the harm of diseases that were once major causes of death for children. We can reflect on the privilege of being in a position to forget certain diseases, but we still need to realize that forgetting can be dangerous. Dr. Perri Klass's *The Best Medicine* travels a complementary road, showing how the successes of medicine and public health changed childhood from a perilous period to one of hope and expectation. There are incredible stories of medical and public health triumphs in our past, and revisiting some of these in classrooms or books or short online videos is a way of not losing sight of our hard-won victories. Giving us amnesia is baked into measles's genetic code, but we can go against nature a bit and use the lessons of measles to create some of our own "memory cells"—by using measles as an example to teach what successful public health campaigns look like, how to communicate with the public, and how to appreciate the tools we have to give children a safer future.

There are less proud moments in public health history, and we must acknowledge those and use them as lessons as well. It's also a way to inspire the next generation of scientists that we will need to step in when new challenges arise. That latter point is also a reason to ensure that our health and science communicators are smart, engaging, and (especially) diverse. There are long-standing issues of underrepresentation in science based on racism and sexism, and the resulting perceptions trickle down to affect how children perceive science or medicine as a potential career. When children are asked to draw a scientist, they are overwhelmingly likely to draw someone male and white. Similar biases pervade medicine. As an example, my own field, pediatrics, is one of those that has traditionally been thought to be "acceptable" for women in medicine. For a very long time, female medical students were steered toward a handful of career options and away from more "male" specialties like surgery. Thankfully, such biases are less common now, and an increasing percentage of surgical trainees are female. Today, well over half of pediatricians starting their careers are female, but, even within pediatrics, considerable inequity remains. Men outnumber women at higher academic ranks and in leadership positions such as department chairs. Black and Latino populations are substantially underrepresented in medicine as well, particularly in leadership roles. Equity and representation among those who enter science, medicine, and public health are important goals in themselves, even more so as we consider whom our children will see as the faces of these fields. The lessons of our successes (and failures) and images of equity should be available and promoted.

Anticipatory guidance, a favorite technique of pediatricians, is another tool that we can use. Pediatricians don't just answer parents' questions (though of course we do that, too), we think about a child's developmental stage and what is coming up next. Childhood, especially infancy, is dynamic—by the time a child's next doctor's visit comes around, they may be doing something new, like walking, that can be a total game-changer in terms of behavior, challenges, and safety. We try to anticipate those next months and proactively address questions and concerns that are likely to arise, sometimes before a parent has considered them. The time to secure furniture to the wall is *before* a child starts pulling up on it. Likewise, we can use the concept of anticipatory guidance to help teach families about how to detect misinformation and pseudoscience before they encounter it. This is what the nurses of EMES were doing when they taught mothers to critically evaluate anti-vaccine claims. It's a cousin of the concept of "pre-bunking" in addition to debunking misinformation, an effective strategy to help counter anti-vaccine messages.

Knowing what comes next is an essential component of anticipatory guidance. Infant development generally follows a predictable path. Disinformation campaigns and epidemics, less so. That's why surveillance is so important as well. I don't mean surveillance in the "Bill Gates is putting microchips in the COVID-19 vaccines so that the government can keep us under surveillance" sense, but in the sense of making sure that our systems for detecting measles, polio, and other infections are as good as they can possibly be. Knowing where outbreaks occur as early as possible is crucial to stopping them. Likewise, understanding where pockets of nonmedical exemptions to school-mandated vaccines exist is an essential component of identifying areas where teaching and clear messaging are needed. Both of these kinds of information gathering are substantially more useful when the data can be collected with fine spatial resolution—think measuring at the level of kilometers rather than counties. As discussed in the previous chapters, outbreaks of measles can occur in small undervaccinated areas hidden in cities and states with high overall vaccine rates. If we look only at the state or city or country level, we miss important information. As the 2015 Disneyland outbreak taught us, there can be vast school-to-school differences in nonmedical exemption rates, and a countywide or statewide percentage may not provide the resolution that is needed to identify and understand these clusters. On the internet, spatial mapping (at least in the sense of physical distance) may be less useful. Several groups are using artificial intelligence strategies to identify clusters of vaccine misinformation online, allowing public health systems to respond as early as possible by deploying accurate information, understanding the claims that are circulating, and providing teaching that is consistent, repeated, and science-based. We can do this at individual levels, too, by calling out anti-vaccine rhetoric—carefully, empathically, but firmly—when we hear it.

Given that we have had a safe and effective vaccine against measles for nearly sixty years, every single measles case that occurs in the world is a massive unforced error. Because it is a bellwether of all sorts of other issues, when measles speaks, we need to listen. Measles thrives on being thought of as inevitable or on small victories being treated as end goals. Clusters of measles cases in places where they have been rare should set off alarm bells. It means that something is being missed. That we are not vaccinating effectively, that anti-vaccine misinformation has spread in a community that is suspicious of traditional public health voices, that children are slipping through the cracks. The response cannot simply be to vaccinate, control the outbreak, and leave. A small outbreak this year is a harbinger of a bigger one a couple of years from now. A cluster of measles in a school is a warning that school mandates are under threat. Responses need to be quick, sustained, sensitive to community needs, and multidisciplinary. The same urgency applies for measles outbreaks in places where they are not rare. It is easy to become numb to the numbers. The waves of measles in low- and middle-income countries that many of us feared at the beginning of the COVID-19 pandemic have now arrived. The involved children need care, of course, but measles spotlights a place where vaccination campaigns, nutrition, and basic services are lacking. Measles is first but probably not last. Our responses in these areas need to be sustained—not just directed at controlling today's outbreaks. Those outbreaks are clues for where to target long-term aid and local capacity building, including restoring or constructing comprehensive health systems.

Finally, the global community needs to commit—immediately, definitively, and with a specific goal date—to the moonshot of measles eradication. Eradication has been a part of the measles discussion since the first days of vaccination in the 1960s. There are a thousand reasons not to do this, and there always will be. And yet it's the right thing to do. I understand—the disease is not even close to under control now. We've lost ground in recent years. We're (still) fighting a different pandemic, we're not done with the polio eradication campaign that was supposed to wrap up in the year 2000, and new viral threats seem to emerge every day. Here is why it's still important.

First, eradicating measles is possible. We have an incredible tool—the measles vaccine. The same one that was developed from the virus that came from a kid at a Massachusetts boarding school in 1954. It works as well now as it did in the 1960s when it was first developed, in the 1980s when it stopped nationwide outbreaks, and in 2016 when we eliminated measles (albeit temporarily) from the Americas. Unlike with influenza, we don't have to change the composition of the vaccine every year. Unlike with tetanus, people don't need boosters every ten years. More than 97 percent of kids who get two doses of a measles-containing vaccine in childhood will have long-lasting protection. In 2010, a WHO global advisory committee concluded that "measles can and should be eradicated." In 2012, the 194 member states of the World Health Assembly endorsed a Global Vaccine Action Plan that included goals of measles elimination in four of the six WHO regions by 2015 and an additional one by 2020. But the global health community has never committed to an eradication date.

Second, we've eradicated diseases before. This is an astonishing achievement—the pinnacle of public health's promise. Smallpox eradication was certified in 1979. It's a different virus, and the specific techniques used in that campaign do not necessarily translate to measles eradication, but smallpox eradication required vision, courage, and resources—like measles eradication would. We've eradicated one infectious disease other than smallpox, and it happens to be measles's closest relative. In 2011, rinderpest virus, the cause of cattle plague, was eradicated after a massive international campaign. If you don't know the rinderpest story, it's one worth hearing. There are lots of great sources, but *This Podcast Will Kill You* has a particularly fun and approachable episode about it.

Third, measles eradication will require an incredible amount of cooperation, interest, capacity-building, and investment. That sounds like a negative, but it's not. Measles eradication is an achievable goal, and it can be a lever by which we can bolster international cooperation. It will cost a lot of money, but it will save a lot of money—cost-benefit analyses suggest a savings of $58 for every $1 spent on measles vaccination programs. We have the most important tools already, but there are technological advances that would smooth the way. Temperature-stable vaccines that eliminated the requirement for constant refrigeration were a crucial innovation that made rinderpest eradication possible. A comparable advance for measles vaccination would aid eradication logistics. Microneedle patches or an oral version of the measles vaccine would eliminate the need for needles, syringes, and trained providers for vaccination, and important work is already being done in these areas. None of these advances is impossible, but all these lines of research need continued support. Committing to this goal will allow us to build infrastructure that will last long after measles is gone. Unlike the smallpox campaign, modern eradication campaigns (including the one for polio) invest in local capacity for both public health and health care systems. These need to be built with local leadership and with an assurance that they will have the resources to remain operational even once eradication is complete. Because effective measles surveillance requires detection of even small clusters of cases, we will need to implement systems that detect pockets of undervaccination and that find and definitively diagnose potential cases as efficiently as possible. These systems, too, would have utility far beyond the eradication date.

Finally, the goal itself is worthy. Measles eradication will save children's lives—millions of them. And we would no longer need to worry about the ebb and flow of funding and interest. Measles kills inequitably, and there is a strong ethical argument for investing the time and money needed to make eradication a reality. There is also a practical aspect—dealing with inequities in vaccine availability is likely the only way that measles eradication will be feasible. This would be inspirational science—the kind that drives innovation and inspires a generation of researchers and public health advocates. Eradication is forever, so this would be a gift to our future. If we build and we teach and we boost as we go along, that gift would last far longer than measles.

When President Franklin Delano Roosevelt wanted to point out inconsistencies in political behavior among his rivals, he was fond of the adage "It all depends on whose baby has the measles." It was a way of saying that people's actions often reveal who has skin in the particular game in question. We must remember that we *all* have skin in this game, and whose baby has the measles is a good indicator for whether we are content with our occasional, fragile, local successes or ready to apply our hard-won knowledge, tools, and memory to the difficult job of making the world both safer and better.

Acknowledgments

This book expresses my own opinions and interpretations, which are not necessarily those of my employers or any other organizations with which I am affiliated. I speak only for myself. Likewise, while I have many people to thank for their expertise, assistance, and encouragement along the way, any errors that remain are mine alone.

I am fortunate to have an extraordinary community of friends, family, colleagues, and mentors—all of whom I appreciate, but only some of whom are listed below. Everything about my life and my work owes a debt to them. Certainly, this book would not exist without their encouragement and patience. I ask for forgiveness from those whom I have omitted.

My agent, Michelle Tessler, took a chance on an unknown author with an idea for a book about measles just as a different pandemic was getting going. She saw the potential for measles to illuminate something about our history and, perhaps, to provide an orthogonal view of the daily challenges we were facing. I am grateful to her for her vision, guidance, and encouragement.

Likewise, the incredible group at Avery understood immediately what I hoped to accomplish with this book and provided the support I needed to get there. I am grateful to Hannah Steigmeyer, Nina Shield, Megan Newman, Tracy Behar, Farin Schlussel, Lindsay Gordon, Casey Maloney, Anne Kosmoski, Katie MacLeod-English, and the rest of the

ACKNOWLEDGMENTS

Avery team. I greatly appreciate the careful eye, depth of knowledge, and thoughtful copyediting of Nancy Inglis.

In researching Peter Panum's journey, I benefited from my correspondence with Dr. Jørgen Mikkelsen, an archivist and senior researcher at the Danish National Archives. Dr. René Flamsholt Christensen provided me with insight into Panum's history and medical training and shared translations of crucial portions of his Panum biography. The staff of the Manuscripts and Archives section of the Yale University Library kindly provided information from the John Franklin Enders archives during a time when the library was closed to outside researchers. I had the privilege of interviewing Dr. Walter Orenstein, Dr. Alan Hinman, and the incredible Blima Marcus, all of whom were generous with their insight and their time.

Dr. Perri Klass is an inspiration, a role model, a friend, and a collaborator. I have benefited from her wisdom, guidance, and humor throughout the writing of this book. Dr. Paul Offit showed me that it was possible to be a doctor, a scientist, an advocate for children's health, and an author. My colleagues Drs. David Oshinsky and Arthur Caplan both encouraged me at a crucial early stage in this process.

My academic families at Yale, Columbia, Children's Hospital of Philadelphia, and New York University have included mentors and friends too numerous to count. My scientific mentors, the physician-scientists Dr. Alice Prince and Dr. Jeffrey Weiser, taught me how to ask a question and then get busy gathering the data to answer it. Dr. Susan Coffin nurtured me during fellowship and helped me make sure that the delicate balance of clinical medicine, research, and family stayed balanced. At NYU, Dr. Catherine Manno has been the rarest of leaders—steady, principled, and kind. For the past decade or so, it has been a privilege to work with the members of the division of pediatric infectious diseases at NYU—a "workfam" like no other. Past and present members of the Ratner Lab have made science an incredibly fun and rewarding journey.

ACKNOWLEDGMENTS

Old friends are the best. Timothy Levin and Rabbi Jonathan Malamy provided ideas, encouragement, and understanding as I put together this work.

I am endlessly grateful to my family—my parents and my siblings and their families have lifted me at every point along the way. Special thanks to Gabriel Rose for allowing me to share the story in the introduction and to Jean-Marc Favreau, whose sage legal advice is surpassed only by his skill with puns.

Shari Gelber and Samantha Ratner are the reason that I have a life filled with stories and medicine and books and adventures and magical pets and side quests and laughter and all the rest. The two of you are absolutely everything.

Notes

INTRODUCTION

xvi **made the disease worse in some:** Steven M. Varga, "Fixing a Failed Vaccine," *Nature Medicine* 15, no. 1 (2009): 21–22, doi: 10.1038/nm0109-21, PMID: 19129777; Maria Florencia Delgado et al., "Lack of Antibody Affinity Maturation Due to Poor Toll-Like Receptor Stimulation Leads to Enhanced Respiratory Syncytial Virus Disease," *Nature Medicine* 15, no. 1 (2009): 34–41, doi: 10.1038/nm.1894, PMID: 19079256.

xvii **"sure can't beat the measles":** Gwynne Hogan, "'Brady Bunch' Episode Fuels Campaigns Against Vaccines—And Marcia's Miffed," NPR, April 28, 2019, https://www.npr.org/sections/health-shots/2019/04/28/717595757/brady-bunch-episode-fuels-campaigns-against-vaccines-and-marcia-s-miffed.

xviii **He'd had a fever:** Note about patient descriptions: Where descriptions of individual patients are included in this book, I have taken care to omit or alter all potentially identifying details while staying true to the overall reason that the patient is discussed. Where names of individuals are used, it is because they have been included in news reports, scientific publications, or other publicly available sources.

xix **hundreds of measles cases:** Jane R. Zucker et al., "Consequences of Undervaccination—Measles Outbreak, New York City, 2018–2019," *New England Journal of Medicine* 382, no. 11 (2020): 1009–17, doi: 10.1056/NEJMoa1912514, PMID: 32160662; Anne A. Gershon et al., "Freedom, Measles, and Freedom from Measles," *New England Journal of Medicine* 382, no. 11 (2020): 983–85, doi: 10.1056/NEJMp2000807, PMID: 32160659.

xxiii **Ebola, COVID-19, mpox, and others:** The mpox nomenclature is still relatively new. In 2022, the World Health Organization and other groups recommended changing the name of this disease, formerly referred to as "monkeypox," in order to diminish racist and stigmatizing language associated with the disease. See World Health Organization, "WHO Recommends New Name for Monkeypox Disease," news release, November 28, 2022, https://www.who.int/news/item/28-11-2022-who-recommends-new-name-for-monkeypox-disease.

xxv **"as clearly defined as the track of a tornado":** Jacob A. Riis, *How the Other Half Lives: Studies Among the Tenements of New York* (New York: Charles Scribner's Sons, 1890).

NOTES

CHAPTER 1: MEET THE MEASLES

4 **virus and host become one:** Chanakha K. Navaratnarajah et al., "Receptor-Mediated Cell Entry of Paramyxoviruses: Mechanisms, and Consequences for Tropism and Pathogenesis," *Journal of Biological Chemistry* 295, no. 9 (2020): 2771–86, doi: 10.1074/jbc.REV119.009961, PMID: 31949044.

4 **these particular proteins fitting together:** Maciej F. Boni, "Evolutionary Origins of the SARS-CoV-2 Sarbecovirus Lineage Responsible for the COVID-19 Pandemic," *Nature Microbiology* 5, no. 11 (2020): 1408–17, doi: 10.1038/s41564-020-0771-4, PMID: 32724171.

5 **their specificity as a harbinger for measles:** Henry E. Koplik, "The Diagnosis of the Invasion of Measles from a Study of the Exanthema as It Appears on the Buccal Mucous Membrane," *Archives of Pediatrics* 13 (1896): 918–22; Henry E. Koplik, "A New Diagnostic Sign of Measles," *Medical Record* 53, no. 15 (1898): 505–507; Howard Markel, "Henry Koplik, MD, the Good Samaritan Dispensary of New York City, and the Description of Koplik's Spots," *Archives of Pediatrics and Adolescent Medicine* 150, no. 5 (1996; 150): 535–39, doi: 10.1001/archpedi.1996.02170300089017, PMID: 8620238.

6 **a different receptor called nectin-4:** Ryan S. Noyce et al., "Tumor Cell Marker PVRL4 (Nectin 4) Is an Epithelial Cell Receptor for Measles Virus," *PloS Pathogens* 7, no. 8 (2011): e1002240, doi: 10.1371/journal.ppat.1002240, PMID: 21901103; Michael D. Mühlebach et al., "Adherens Junction Protein Nectin-4 Is the Epithelial Receptor for Measles Virus," *Nature* 480, no. 7378 (2011): 530–33, doi: 10.1038/nature10639, PMID: 22048310.

6 **virions and even clusters of whole infected cells:** Camilla E. Hippee et al., "Measles Virus Exits Human Airway Epithelia Within Dislodged Metabolically Active Infectious Centers," *PloS Pathogens* 17, no. 8 (2021): e1009458, doi: 10.1371/journal.ppat.1009458, PMID: 34383863.

7 **"almost as contagious as measles":** Perri Klass and Adam J. Ratner, "'The Sombre Aspect of the Entire Landscape'—Epidemiology and the Faroe Islands," *New England Journal of Medicine* 386, no. 13 (2022): 1202–205, doi: 10.1056/NEJMp2120194, PMID: 35333484.

CHAPTER 2: STORYTELLING

12 **died of COVID-19 in February 2020:** Muyi Xiao, Isabelle Qian, Tracy Wen Liu, and Chris Buckley, "How a Chinese Doctor Who Warned of Covid-19 Spent His Final Days," *New York Times*, October 6, 2022, https://www.nytimes.com/2022/10/06/world/asia/covid-china-doctor-li-wenliang.html.

15 **wrote a short paper:** Michal Paret et al., "Severe Acute Respiratory Syndrome Coronavirus 2 (SARS-CoV-2) Infection in Febrile Infants Without Respiratory Distress," *Clinical Infectious Diseases* 71, no. 16 (2020): 2243–45, doi: 10.1093/cid/ciaa452, PMID: 32301967.

CHAPTER 3: PANUM'S MAP

17 **Thorshavn:** I have generally used the names and spellings of Faroese places that are used in Panum's writings. In some cases, these differ from modern usage, and in some cases multiple terms are still used. Tórshavn is the common spelling of this city today.

NOTES

17 **In late March 1846:** The major source for this chapter is Peter Ludvig Panum, *Observations Made During the Epidemic of Measles on the Faroe Islands in the Year 1846: With a Biographical Memoir . . . and an Introduction* (New York: Delta Omega Society; distributed by the American Public Health Association, 1940), with a biographical memoir by Julius Jacob Petersen and an introduction by James Angus Doull. Holm was identified in a biography of Panum: René Flamsholt Christensen, *Peter Ludvig Panum: det moderne gennembrud i dansk medicin* (Roskilde: FADL, 2020).

18 **a fort without a cannon:** James Nicol, *An Historical and Descriptive Account of Iceland, Greenland, and the Faroe Islands* (Edinburgh: Oliver & Boyd, 1840).

19 **Candidate August Manicus:** Norbert B. Vogt and Uwe Kordek, "Works in English From and About the Faroe Islands: An Annotated Bibliography," Fróðskaparrit 47 (1999): 33–127; "Læge Manicus' beretning fra Færøerne om mæslingeepidemien i 1846" [Dr. Manicus's report from the Faroe Islands on the measles epidemic in 1846], danmarkshistorien.dk (Aarhus University), https://danmarkshistorien.dk/vis/materiale/uddrag-af-laege-andreas-heinrich-manicus-beretning-fra-faeroeerne-i-forbindelse-med-maeslingeepidemien/.

19 **Candidate Peter Ludvig Panum:** Gill Roper-Hall and Heidi Jørgensen, "Historical Vignette: Peter Ludvig Panum (1820–1885), Danish Physician and Physiologist," *American Orthoptic Journal* 58 (2008): 99–107, doi: 10.3368/aoj.58.1.99, PMID: 21149184.

19 **little in Panum's history:** Flamsholt Christensen, *Peter Ludvig Panum*.

19 **rampant elsewhere in the Faroes:** Panum, *Observations Made During the Epidemic of Measles on the Faroe Islands in the Year 1846;* "Læge Manicus' beretning fra Færøerne om mæslingeepidemien i 1846 [Dr. Manicus's report from the Faroe Islands on the measles epidemic in 1846].

21 **advanced the concept of *seminaria contagiosa*:** Charles-Edward Amory Winslow, *The Conquest of Epidemic Disease: A Chapter in the History of Ideas* (Madison: University of Wisconsin Press, 1980).

22 **nearly six births out of a thousand:** Irvine Loudon, "Deaths in Childbed from the Eighteenth Century to 1935," *Medical History* 30, no. 1 (1986): 1–41, doi: 10.1017/s0025727300045014, PMID: 3511335.

22 **"not as a misfortune but as a crime":** Oliver Wendell Holmes, "The Contagiousness of Puerperal Fever," *New England Quarterly Journal of Medicine and Surgery* 1843, https://archive.org/details/67241030R.nlm.nih.gov.

22 **simple hygienic interventions could prevent:** Sherwin B. Nuland, *The Doctors' Plague: Germs, Childbed Fever, and the Strange Story of Ignác Semmelweis* (New York: Norton, 2004).

22 **the miasma hypothesis was faltering:** Winslow, *The Conquest of Epidemic Disease*.

23 ***Treatise on Smallpox and Measles*:** Abú Becr Mohammed ibn Zacaríyá ar-Rází (commonly called Rhazes), *A Treatise on the Small-pox and Measles*, translated from the original Arabic by William Alexander Greenhill (London: Sydenham Society, 1748), https://iiif.wellcomecollection.org/pdf/b29341073.

23 **doctors needed to be as observant as artists:** Thomas Sydenham, *The Works of Thomas Sydenham, M.D.*, translated from the Latin edition of Dr. Greenhill, with a life of the

NOTES

author, by R. G. Latham. Epistle dedication to the third edition (London: Sydenham Society, 1848), https://www.google.com/books/edition/The_Works_of_Thomas_Sydenham_M_D/8qYEAAAAYAAJ?hl=en&gbpv=1.

25 **An account from the 1800s:** Nicol, *An Historical and Descriptive Account of Iceland, Greenland, and the Faroe Islands.*

28 **a contemporary medical textbook:** Marshall Hall, Jacob Bigelow, and Oliver Wendell Holmes, *Principles of the Theory and Practice of Medicine* (Boston: Charles C. Little and James Brown, 1839), https://www.google.com/books/edition/Principles_of_the_Theory_and_Practice_of/9edMAQAAMAAJ?hl=en&gbpv=0.

28 **"When a physician is called to work":** Peter Ludwig Panum, "Iagttagelser, anstillede under Maeslinge-Epidemien paa Faeroerne i Aaret 1846" [Observations made during the measles epidemic on the Faroes in the year 1846], *Bibliotek for Laeger* 3 (1847): 270–344; translation in *Medical Classics* 3 (1939): 829–86.

31 **completely unknown to the residents:** Panum appears to be wrong on this point, at least as pertains to whooping cough. See Danish medical statistics, *Boston Medical and Surgical Journal* 23 (1840): 324–28, especially 327, doi: 10.1056/NEJM184012230232004.

33 **use quarantine and isolation systematically:** Hermann Nothnagel and Alfred Stengel, *Nothnagel's Encyclopedia of Practical Medicine* (Philadelphia: Saunders, 1902), 228–29.

34 **a hand-drawn map of the Faroes:** Rigsarkivet, *Sundhedsstyrelsen, Journalsager 1806–1981*, pakke 120 (sag 11/1847). Image courtesy of Jørgen Mikkelsen, PhD, archivist and senior researcher, the Danish National Archives. Available at "Læge Manicus' beretning fra Færøerne om mæslingeepidemien i 1846" [Dr. Manicus's report from the Faroe Islands on the measles epidemic in 1846].

34 **twin articles on their experiences:** Panum, "Iagttagelser, anstillede under Maeslinge-Epidemien paa Faeroerne i Aaret 1846" [Observations made during the measles epidemic on the Faroes in the year 1846]; A. Manicus, "Maeslingerne paa Faeröerne i Sommeren 1846," *Ugeskrift for Laeger* 6 (1847): 189–210, https://www.google.com/books/edition/Ugeskrift_for_læger/MOoCAAAAYAAJ.

35 **spread of the omicron variant of SARS-CoV-2 within gatherings of fully vaccinated people:** Gunnhild Helmsdal et al., "Omicron Outbreak at a Private Gathering in the Faroe Islands, Infecting 21 of 33 Triple-Vaccinated Healthcare Workers," *Clinical Infectious Diseases* 75, no. 5 (2022): 893–96, doi: 10.1093/cid/ciac089, PMID: 35134167; Perri Klass and Adam J. Ratner, "'The Sombre Aspect of the Entire Landscape'—Epidemiology and the Faroe Islands," *New England Journal of Medicine* 386, no. 13 (2022): 1202–205, doi: 10.1056/NEJMp2120194, PMID: 35333484.

35 **"cholera map" investigations:** Steven Johnson, *The Ghost Map: The Story of London's Most Terrifying Epidemic—And How It Changed Science, Cities, and the Modern World* (New York: Riverhead Books, 2006).

36 **empire and bureaucracy and colonial exploitation:** Jim Downs, *Maladies of Empire: How Colonialism, Slavery, and War Transformed Medicine* (Cambridge, MA: Belknap Press of Harvard University Press, 2021).

NOTES

CHAPTER 4: CONTAGION IN THE SERVICE OF EMPIRE

39 **"civilized diseases":** William H. McNeill, *Plagues and Peoples* (New York: Anchor Press/Doubleday, 1976).

40 **smallpox, measles, influenza:** Jared Diamond, *Guns, Germs, and Steel: The Fates of Human Societies* (New York: Norton, 2017).

40 **less consistently on the guns and steel:** Noble David Cook, *Born to Die: Disease and New World Conquest, 1492–1650* (New York: Cambridge University Press, 1998); Charles C. Mann, *1491: New Revelations of the Americas Before Columbus* (New York: Knopf, 2005).

41 **incomplete or flat-out wrong:** David S. Jones, "Virgin Soils Revisited," *William & Mary Quarterly* 60, no. 4 (2003): 703–42, doi:10.2307/3491697.

42 **multiple overlapping outbreaks:** Alfred W. Crosby Jr., *The Columbian Exchange: Biological and Cultural Consequences of 1492* (Westport, CT: Greenwood Press, 1972), 47.

42 **measles had definitively joined smallpox:** Cook, *Born to Die: Disease and New World Conquest, 1492–1650*, 86.

43 **"foreign pathogens were active":** Cook, *Born to Die: Disease and New World Conquest, 1492–1650*, 95.

44 **unfathomable loss:** Mann, *1491: New Revelations of the Americas Before Columbus*, 144.

44 **situating them in a "death trap":** Cook, *Born to Die: Disease and New World Conquest, 1492–1650*, 123.

44 **"ran ahead of direct contact":** McNeill, *Plagues and Peoples*, 210; Diamond, *Guns, Germs, and Steel: The Fates of Human Societies*, 373–74; Cook, *Born to Die: Disease and New World Conquest, 1492–1650*, 162.

44 **first among a long list of divine gifts:** New England's First Fruits, *Collections of the Massachusetts Historical Society*, 1643; series 1, vol. I, 242–50 (Boston, MA: Munroe & Francis), 246; Herbert Upham Williams, "The Epidemic of the Indians of New England, 1616–1620, with Remarks on Native American Infections," *Johns Hopkins Hospital Bulletin* 20, no. 234 (1909): 340–49.

44 **preaching and engaging in public discourse:** Paul E. Kopperman and Jeanne Abrams, "Cotton Mather's Medicine, with Particular Reference to Measles," *Journal of Medical Biography* 27, no. 1 (2019): 30–37, doi: 10.1177/0967772016662166, PMID: 27635032.

45 **"to make room for a better growth":** Cotton Mather, *Magnalia Christi Americana, Or the Ecclesiastical History of New England from 1620–1698*, book I, chapter 2, 7, https://www.google.com/books/edition/Magnalia_Christi_Americana/GNBDAAAAcAAJ?hl=en&gbpv=0.

45 **died during the outbreak:** Kopperman and Abrams, "Cotton Mather's Medicine, with Particular Reference to Measles." Original citation is David Levin, *Cotton Mather: The Young Life of the Lord's Remembrancer, 1663–1703* (Cambridge, MA: Harvard University Press, 1978), 304–307.

45 **"common Calamity of the spreading measles":** Diary of Cotton Mather, vol. 2, 18 October 1713, https://catalog.hathitrust.org/Record/100323459.

46 **citywide days of fasting and prayer:** E. Caulfield, "Early Measles Epidemics in America," *Yale Journal of Biology and Medicine* 15, no. 4 (1943): 531–56, PMID: 21434087.

NOTES

46 **"The Measles are a Distemper":** Cotton Mather, "A LETTER, About a Good Management under the Distemper of the MEASLES, at this time Spreading in the Country. Here Published for the Benefit of the Poor, and such as may want the help of Able Physicians," https://quod.lib.umich.edu/e/evans/N29746.0001.001.

47 **a progressively less susceptible population:** Caulfield, "Early Measles Epidemics in America."

48 **traveled to Sydney for a state visit:** William Squire, "On Measles in Fiji," *Transactions: Epidemiological Society of London* 4, pt. 1 (1879): 72–74, PMID: 29418970; Bolton G. Corney, "The Behaviour of Certain Epidemic Diseases in Natives of Polynesia, with Especial Reference to the Fiji Islands," *Transactions: Epidemiological Society of London* 3 (1884): 76–95, PMID: 29419050; D. M. Morens, "Measles in Fiji, 1875: Thoughts on the History of Emerging Infectious Diseases," *Pacific Health Dialog* 5 (1981): 119–28; A. D. Cliff, Peter Haggett, and Matthew Smallman-Raynor, *Measles: An Historical Geography of a Major Human Viral Disease from Global Expansion to Local Retreat, 1840–1990* (Oxford, UK: Blackwell, 1993), 130–33; R. A. Derrick, "1875: Fiji's Darkest Hour—An Account of the Measles Epidemic of 1875," *Transactions and Proceedings of the Fiji Society for the Years 1955–1957* 6, no. 1 (1955), 3–16, http://www.justpacific.com/fiji/full-text/Derrick%E2%80%94Measles.pdf; G. Dennis Shanks, "Pacific Island Societies Destabilized by Infectious Diseases," *Journal of Military and Veterans' Health* 24 (2016): 71–74.

48 **massive measles outbreak in Eastern Australia:** "Measles Epidemics in Victoria," Museums Victoria Collections, https://collections.museumsvictoria.com.au/articles/16826.

48 **death rates from measles were even higher:** Peter John Dowling, *"A Great Deal of Sickness": Introduced Diseases Among the Aboriginal People of Colonial Southeast Australia, 1788–1900*, PhD dissertation, Order 9936134, Australian National University (Australia), 1998, https://www.proquest.com/docview/304465029/fulltextPDF/D0A1186E780340B2PQ/1?; Peter John Dowling, *Fatal Contact: How Epidemics Nearly Wiped Out Australia's First Peoples* (Clayton, Victoria: Monash University Publishing, 2021).

48 **infected during his state visit:** There is ambiguity on this point: Cliff, Haggett, and Smallman-Raynor, *Measles: An Historical Geography of a Major Human Viral Disease from Global Expansion to Local Retreat, 1840–1990*, 121.

48 **looked weak on arrival:** *Fiji Times* (Lekuva), December 13, 1874—cited in Cliff, Haggett, and Smallman-Raynor, *Measles: An Historical Geography of a Major Human Viral Disease from Global Expansion to Local Retreat, 1840–1990*.

48 **ship's physician diagnosed:** Accounts differ—see Derrick, "1875: Fiji's Darkest Hour—An Account of the Measles Epidemic of 1875," 5; Corney, "The Behaviour of Certain Epidemic Diseases in Natives of Polynesia, with Especial Reference to the Fiji Islands," 78.

49 **Cases and deaths were widespread:** Corney, "The Behaviour of Certain Epidemic Diseases in Natives of Polynesia, with Especial Reference to the Fiji Islands," 80.

50 **"decapitation of society":** Shanks, "Pacific Island Societies Destabilized by Infectious Diseases," 72.

50 **"first advantage derived from annexation":** *Fiji Times*, February 17, 1875—cited in Derrick, "1875: Fiji's Darkest Hour—An Account of the Measles Epidemic of 1875," 5;

NOTES

Corney, "The Behaviour of Certain Epidemic Diseases in Natives of Polynesia, with Especial Reference to the Fiji Islands," 8.

51 **more focused decapitation:** Stanford T. Shulman, Deborah L. Shulman, and Ronald H. Sims, "The Tragic 1824 Journey of the Hawaiian King and Queen to London: History of Measles in Hawaii," *Pediatric Infectious Disease Journal* 28, no. 8 (2009): 728–33, doi: 10.1097/INF.0b013e31819c9720, PMID: 19633516; Robert C. Schmitt and Eleanor C. Nordyke, "Death in Hawai'i: The Epidemics of 1848–1849," *Hawaiian Journal of History* 35 (2001): 1–13, https://evols.library.manoa.hawaii.edu/bitstream/handle/10524/339/JL35007.pdf?sequence=2.

51 **initial introduction of measles:** Shulman, Shulman, and Sims, "The Tragic 1824 Journey of the Hawaiian King and Queen to London: History of Measles in Hawaii"; Schmitt and Nordyke, "Death in Hawai'i: The Epidemics of 1848–1849."

51 **joining the "global pathogen pool":** Morens, "Measles in Fiji, 1875: Thoughts on the History of Emerging Infectious Diseases"; Andrew Cliff and Peter Haggett, "Time, Travel and Infection," *British Medical Bulletin* 69 (2004): 87–99, doi: 10.1093/bmb/ldh011, PMID: 15226199.

51 **Such trends accelerated:** Cliff and Haggett, "Time, Travel and Infection."

CHAPTER 5: CROWDED, POOR, MALNOURISHED

53 **more than fifty thousand people seeking asylum:** Ashwin Vasan, letter to colleagues, April 11, 2023, https://www.nyc.gov/assets/doh/downloads/pdf/immigrant-health/asylum-seeker-dear-colleague.pdf.

54 **These include:** Rory D. de Vries, W. Paul Duprex, and Rik L. de Swart, "Morbillivirus Infections: An Introduction," *Viruses* 7, no. 2 (2015): 699–706, doi: 10.3390/v7020699, PMID: 25685949.

55 **revealed ancient human migration patterns:** Daniel Falush et al., "Traces of Human Migrations in *Helicobacter pylori* Populations," *Science* 299, no. 5612 (2003): 1582–85, doi: 10.1126/science.1080857, PMID: 12624269.

55 **ancient genomes of *Yersinia pestis*:** Maria A. Spyrou et al., "Analysis of 3800-Year-Old *Yersinia pestis* Genomes Suggests Bronze Age Origin for Bubonic Plague," *Nature Communications* 9, no. 1 (2018): 2234, doi: 10.1038/s41467-018-04550-9, PMID: 29884871.

55 **emergence and spread of Ebola:** Stephen K. Gire et al., "Genomic Surveillance Elucidates Ebola Virus Origin and Transmission During the 2014 Outbreak," *Science* 345, no. 6202 (2014): 1369–72, doi: 10.1126/science.1259657, PMID: 25214632.

56 **spillover of SARS-CoV-2 from bats:** Maciej F. Boni et al., "Evolutionary Origins of the SARS-CoV-2 Sarbecovirus Lineage Responsible for the COVID-19 Pandemic," *Nature Microbiology* 5, no. 11 (2020): 1408–17, doi: 10.1038/s41564-020-0771-4, PMID: 32724171.

56 **separation of measles and rinderpest viruses:** Murali Muniraju et al., "Molecular Evolution of Peste des Petits Ruminants Virus," *Emerging Infectious Diseases* 20, no. 12 (2014): 2023–33, doi: 10.3201/eid2012.140684, PMID: 25418782; Hirokazu Kimura et al.,

NOTES

"Molecular Evolution of Haemagglutinin (H) Gene in Measles Virus," *Scientific Reports* 5 (2015): 11648, doi: 10.1038/srep11648, PMID: 26130388.

56 **implicated the transatlantic slave trade:** Axel A. Guzmán-Solís et al., "Ancient Viral Genomes Reveal Introduction of Human Pathogenic Viruses into Mexico During the Transatlantic Slave Trade," *eLife* 10 (2021): e68612, PMID: 34350829.

56 **a corpse recovered from Alaskan permafrost:** Ann H. Reid et al., "Origin and Evolution of the 1918 'Spanish' Influenza Virus Hemagglutinin Gene," *Proceedings of the National Academy of Sciences of the USA* 96, no. 4 (1999): 1651–56, doi: 10.1073/pnas.96.4.1651, PMID: 9990079; Jeffery K. Taubenberger et al., "Initial Genetic Characterization of the 1918 'Spanish' Influenza Virus," *Science* 275, no. 5307 (1997): 1793–96, doi: 10.1126/science.275.5307.1793, PMID: 9065404.

57 **nearly half a century earlier:** Jeffery K. Taubenberger, Johan V. Hultin, and David M. Morens, "Discovery and Characterization of the 1918 Pandemic Influenza Virus in Historical Context," *Antiviral Therapy* 12, no. 4, pt. B (2007): 581–91, PMID: 17944266; Elizabeth Fernandez, "The Virus Detective / Dr. John Hultin Has Found Evidence of the 1918 Flu Epidemic That Had Eluded Experts for Decades," SF Gate, February 17, 2002, https://www.sfgate.com/magazine/article/The-Virus-detective-Dr-John-Hultin-has-found-2872017.php; Jeffery K. Taubenberger, John C. Kash, and David M. Morens, "The 1918 Influenza Pandemic: 100 Years of Questions Answered and Unanswered," *Science Translational Medicine* 11, no. 502 (2019): eaau5485, doi: 10.1126/scitranslmed.aau5485, PMID: 31341062.

57 **"find bodies in the permafrost":** Gina Kolata, *Flu: The Story of the Great Influenza Pandemic of 1918 and the Search for the Virus That Caused It* (New York: Touchstone, 1999); Gina Kolata, "Johan Hultin, Who Found Frozen Clues to 1918 Virus, Dies at 97," January 27, 2022, https://www.nytimes.com/2022/01/27/health/dr-johan-hultin-dead.html; Taubenberger, Hultin, and Morens, "Discovery and Characterization of the 1918 Pandemic Influenza Virus in Historical Context."

57 **journeyed to Brevig Mission:** Alfred W. Crosby, *America's Forgotten Pandemic: The Influenza of 1918*, 2nd ed. (Cambridge: Cambridge University Press, 2003), 305–306.

57 **discovered the body of a young girl:** Kolata, *Flu*.

58 **closely related to modern influenza strains:** Taubenberger et al., "Initial Genetic Characterization of the 1918 'Spanish' Influenza Virus"; Taubenberger, Kash, and Morens, "The 1918 Influenza Pandemic: 100 Years of Questions Answered and Unanswered."

58 **self-financed return to Alaska:** Fernandez, "The Virus Detective / Dr. John Hultin Has Found Evidence of the 1918 Flu Epidemic That Had Eluded Experts for Decades."

59 **clarify the evolutionary origin:** Tuofu Zhu et al., "An African HIV-1 Sequence from 1959 and Implications for the Origin of the Epidemic," *Nature* 391, no. 6667 (1998): 594–97, doi: 10.1038/35400, PMID: 9468138; Michael Worobey et al., "Direct Evidence of Extensive Diversity of HIV-1 in Kinshasa by 1960," *Nature* 455, no. 7213 (2008): 661–64, doi: 10.1038/nature07390, PMID: 18833279; Sophie Gryseels et al., "A Near Full-Length HIV-1 Genome from 1966 Recovered from Formalin-Fixed Paraffin-Embedded Tissue,"

NOTES

Proceedings of the National Academy of Sciences of the USA 117, no. 22 (2020): 12222–229, doi: 10.1073/pnas.1913682117, PMID: 32430331.

59 **another recent success:** Ariane Düx et al., "Measles Virus and Rinderpest Virus Divergence Dated to the Sixth Century BCE," *Science* 368, no. 6497 (2020): 1367–70, doi: 10.1126/science.aba9411, PMID: 32554594; Sarah Zhang, "The Virus Buried in a 100-Year-Old Lung," *The Atlantic*, January 9, 2020, https://www.theatlantic.com/science/archive/2020/01/what-100-year-old-lung-says-about-measles-origin/604591.

59 **Childhood deaths are difficult to process:** Perri Klass, *The Best Medicine: How Science and Public Health Gave Children a Future* (New York: Norton, 2022); Perri Klass and Adam J. Ratner, "'The Sombre Aspect of the Entire Landscape'—Epidemiology and the Faroe Islands," *New England Journal of Medicine* 386, no. 13 (2022): 1202–205, doi: 10.1056/NEJMp2120194, PMID: 35333484.

60 **identified the girl's preserved lung:** Zhang, "The Virus Buried in a 100-Year-Old Lung."

61 **that population threshold:** M. S. Bartlett, "Measles Periodicity and Community Size," *Journal of the Royal Statistical Society: Series A (General)* 120, no. 1 (1957): 48–70, https://doi.org/10.2307/2342553; M. S. Bartlett, "The Critical Community Size for Measles in the United States," *Journal of the Royal Statistical Society: Series A (General)* 123, no. 1 (1960): 37–44, https://doi.org/10.2307/2343186; D. Cliff, Peter Haggett, and Matthew Smallman-Raynor, *Measles: An Historical Geography of a Major Human Viral Disease from Global Expansion to Local Retreat, 1840–1990* (Oxford, UK: Blackwell, 1993); Francis L. Black, "Measles Endemicity in Insular Populations: Critical Community Size and Its Evolutionary Implication," *Journal of Theoretical Biology* 11, no. 2 (1966): 207–11, doi: 10.1016/0022-5193(66)90161-5, PMID: 5965486; M. J. Keeling and B. T. Grenfell, "Disease Extinction and Community Size: Modeling the Persistence of Measles," *Science* 275, no. 5296 (1997): 65–67, doi: 10.1126/science.275.5296.65, PMID: 8974392.

61 **unprecedented densities:** Serge Morand, K. Marie McIntyre, and Matthew Baylis, "Domesticated Animals and Human Infectious Diseases of Zoonotic Origins: Domestication Time Matters," *Infection, Genetics, and Evolution*, 24 (2014): 76–81, doi: 10.1016/j.meegid.2014.02.013, PMID: 24642136.

61 **enhancing the potential for pathogen spillover:** William H. McNeill, *Plagues and Peoples* (New York: Anchor Press/Doubleday, 1976), 47.

61 **multiple communities crossing the threshold:** Hiroko Inoue et al., "Urban Scale Shifts Since the Bronze Age: Upsweeps, Collapses, and Semiperipheral Development," *Social Science History* 39, no. 2 (2015): 175–200, https://www.jstor.org/stable/90017172; Düx et al., "Measles Virus and Rinderpest Virus Divergence Dated to the Sixth Century BCE."

62 **at least four Nipah outbreaks:** World Health Organization, "Disease Outbreak News: Nipah Virus Infection—India," October 3, 2023, https://www.who.int/emergencies/disease-outbreak-news/item/2023-DON490.

62 **London had become measles endemic:** Charles Creighton, *A History of Epidemics in Britain*, vol. 2 (Cambridge: Cambridge University Press, 1894), 643, https://www

NOTES

.google.com/books/edition/A_History_of_Epidemics_in_Britain/7XsaAAAAMAAJ?hl=en&gbpv=0.

62 **hundreds of measles deaths every year:** Cliff, Haggett, and Smallman-Raynor, *Measles: An Historical Geography of a Major Human Viral Disease from Global Expansion to Local Retreat, 1840–1990*, 98–101; John Duffy, *A History of Public Health in New York City, 1625–1866* (New York: Russell Sage Foundation, 1974), https://www.russellsage.org/publications/history-public-health-new-york-city-1625-1866.

62 **"everyone must have measles":** W. Butler, "Measles," *Proceedings of the Royal Society of Medicine* 6 (1913): 120–37, PMID: 19977230.

63 **concentrated in young children:** G. D. Shanks et al., "Age-Specific Measles Mortality During the Late 19th–Early 20th Centuries," *Epidemiology & Infection* 143, no. 16 (2015): 3434–41, doi: 10.1017/S0950268815000631, PMID: 25865777; C. O'Donovan, "Measles in Kenyan Children," *East African Medical Journal* 48, no. 10 (1971): 526–32, PMID: 5141408.

63 **"a necessary part of the game":** Anne Hardy, *The Epidemic Streets: Infectious Disease and the Rise of Preventive Medicine, 1856–1900* (Oxford: Clarendon Press, 1993), table 2.1.

64 **"every man must catch measles":** F. J. Waldo and D. Walsh, "Murder by Measles," *The Nineteenth Century: A Monthly Review, March 1877–December 1900*, 39 (June 1896): 957–963, https://www.google.com/books/edition/The_Nineteenth_Century/PMRMAQAAMAAJ?hl=en-.

64 **"may I have the measles when":** A. T. Smith, "Tommy (watching delicacies being taken to invalid). 'Mummy, may I have the measles when Violet's finished with them?'" *Punch* 162, no. 4237 (February 22, 1922): 145, Punch Historical Archive, 1841–1992, https://link.gale.com/apps/doc/ES700026288/PNCH?u=new64731&sid=PNCH&xid=23be6a5b.

64 **"the great poverty of those residing therein":** Harry Drinkwater, *Remarks upon the Epidemic of Measles Prevalent in Sunderland: With Notes upon 311 Cases from Middle of January to End of March 1885*, Edinburgh Medical School thesis and dissertation collection, March 1885, 8, https://www.google.com/books/edition/Remarks_Upon_the_Epidemic_of_Measles_Pre/ATQAAAAAQAAJ?hl=en&gbpv=0.

65 **"a malady that calls for little or no treatment":** Waldo and Walsh, "Murder by Measles."

65 **vitamin A deficiency as a particularly potent risk factor for severe measles:** National Foundation for Infectious Diseases, "Call to Action: Vitamin A for the Management of Measles in the United States," 2020, https://www.nfid.org/resource/vitamin-a-for-the-management-of-measles-in-the-us/.

66 **importance of socioeconomic disparities in health:** Hardy, *The Epidemic Streets: Infectious Disease and the Rise of Preventive Medicine, 1856–1900*; R. M. Picken, "The Epidemiology of Measles in a Rural and Residential Area," *Proceedings of the Royal Society of Medicine* 14 (1921): 75–84, PMID: 19981949.

66 **"The outstanding characteristic of a tenement building":** J. L. Halliday, "An Inquiry into the Relationship Between Housing Conditions and the Incidence and Fatality of Measles," MRC report series special no. 120, May 22, 1928, https://www.cabdirect.org/cabdirect/abstract/19282701675.

NOTES

67 **"made proper care of the sick impossible":** Jacob A. Riis, *How the Other Half Lives* (New York: Charles Scribner's Sons, 1890), https://www.gutenberg.org/files/45502/45502-h/45502-h.htm.

67 **intensity of viral exposure:** Peter Aaby et al., "High Measles Mortality in Infancy Related to Intensity of Exposure," *Journal of Pediatrics* 109, no. 1 (1986): 40–44, doi: 10.1016/s0022-3476(86)80569-8, PMID: 3723239; Peter Aaby et al., "Severe Measles in Sunderland, 1885: A European-African Comparison of Causes of Severe Infection," *International Journal of Epidemiology* 15, no. 1 (1986): 101–107, doi: 10.1093/ije/15.1.101, PMID: 3957529; Peter Aaby et al., "Determinants of Measles Mortality in a Rural Area of Guinea-Bissau: Crowding, Age, and Malnutrition," *Journal of Tropical Pediatrics* 30, no. 3 (1984): 164–68, doi: 10.1093/tropej/30.3.164, PMID: 6737555; O'Donovan, "Measles in Kenyan Children."

68 **nearly 11 percent:** J. K. Barnes et al., *Medical and Surgical History of the War of the Rebellion* (Washington, DC: Government Publishing Office, 1870), https://collections.nlm.nih.gov/catalog/nlm:nlmuid-14121350R-mvset, cited in Matthew Smallman-Raynor and Cliff Andrew, *War Epidemics: An Historical Geography of Infectious Diseases in Military Conflict and Civil Strife, 1850–2000* (Oxford: Oxford University Press, 2006), 188.

68 **because of malnutrition and inhumane living conditions:** Susan E. Klepp, "Seasoning and Society: Racial Differences in Mortality in Eighteenth-Century Philadelphia," *William & Mary Quarterly* 51, no. 3 (1994): 473–506, https://doi.org/10.2307/2947439.

69 **"That dark bedroom killed it":** Riis, *How the Other Half Lives*.

CHAPTER 6: MAKING NOTHING HAPPEN

72 **a progressive and fatal neurological disease:** William J. Bellini et al., "Subacute Sclerosing Panencephalitis: More Cases of This Fatal Disease Are Prevented by Measles Immunization Than Was Previously Recognized," *Journal of Infectious Diseases* 192, no. 10 (2005): 1686–93, doi: 10.1086/497169, PMID: 16235165.

72 **Protection through inoculation:** Arthur Boylston, "The Origins of Inoculation," *Journal of the Royal Society of Medicine* 105, no. 7 (2012): 309–13, doi: 10.1258/jrsm.2012.12k044, PMID: 22843649.

73 **a procedure called insufflation:** Boylston, "The Origins of Inoculation."

73 **method for smallpox prevention:** Emanuel Timonius, "V. An Account, or History, of the Procuring the Smallpox by Incision, or Inoculation; As It Has for Some Time Been Practised at Constantinople," *Philosophical Transactions of the Royal Society* 29 (1714): 72–82, doi: 10.1098/rstl.1714.0010.

73 **as a "gift" from his congregation:** Diary of Cotton Mather, vol. 1, 579, https://catalog.hathitrust.org/Record/100323459.

74 **"in his Arm the scar":** G. L. Kittredge, "Some Lost Works of Cotton Mather," *Proceedings of the Massachusetts Historical Society* 45 (1911–1912), third series: 418–79, https://www.jstor.org/stable/25080004, 422.

74 **"Cotton Mather, You Dog, Dam you":** Diary of Cotton Mather, vol. 2, 658; also see Reginald H. Fitz, "Zabdiel Boylston, Inoculator, and the Epidemic of Smallpox in Boston in 1721," *Bulletin of the Johns Hopkins Hospital* 22, no. 247 (1911): 315–27.

NOTES

74 **had it performed on her young son:** Lewis Melville and Lady Mary Wortley, *Lady Mary Wortley Montague, Her Life and Letters (1689–1762)*, https://www.gutenberg.org/cache/epub/10590/pg10590-images.html.

74 **she promoted the procedure in England:** Diana Barnes, "The Public Life of a Woman of Wit and Quality: Lady Mary Wortley Montagu and the Vogue for Smallpox Inoculation," *Feminist Studies* 38, no. 2 (2012): 330–62, http://www.jstor.org/stable/23269190.

75 **Dobson reportedly saved:** *The Gentleman's Magazine*: and Historical Chronicle, January 1736–December 1833, London, vol. 24 (1754), 493. Of note, the letter was disputed by Dobson the next month. Dobson had heard about the procedure from someone else and, as a city dweller, reported that he did not even keep a cow. *Gentleman's Magazine*, December 1754, 549. The discrepancy is noted in C. Huygelen, "The Immunization of Cattle Against Rinderpest in Eighteenth-Century Europe," *Medical History* 41, no. 2 (1997): 182–96, doi: 10.1017/s0025727300062372, PMID: 9156464.

75 **measles, and potentially other diseases:** C. Brown, *Dissertatio medica inauguralis de morbillis* (Edinburgh: Hamilton and Balfour, 1775), 3; Huygelen, "The Immunization of Cattle Against Rinderpest in Eighteenth-Century Europe."

75 **took the bait:** J. F. Enders, "Francis Home and His Experimental Approach to Medicine," *Bulletin of the History of Medicine* 38, no. 2 (1964): 101–12, PMID: 14132122; Stanley A. Plotkin, "Vaccination Against Measles in the 18th Century," *Clinical Pediatrics* (Philadelphia) 6, no. 5 (1967): 312–15; doi: 10.1177/000992286700600524; C. Huygelen, "The Long Prehistory of Modern Measles," in Stanley A. Plotkin, ed., *History of Vaccine Development* (New York: Springer, 2011); W. E. Home, "Francis Home (1719–1813), First Professor of Materia Medica in Edinburgh," *Proceedings of the Royal Society of Medicine* 21, no. 6 (1928): 1013–15, PMID: 19986439.

75 **"no small service to mankind":** Francis Home, *Medical Facts and Experiments* (London, 1759), 266, https://www.google.com/books/edition/Medical_Facts_and_Experiments/-tVhAAAAcAAJ?hl=en&gbpv=0.

76 **no further trials:** Francis Home, *Principia medicae*, cited in Huygelen, "The Long Prehistory of Modern Measles."

77 **from enthusiastic to dismissive:** J. Cooke, "Of Inoculating the Measles," *Universal Magazine of Knowledge and Pleasure* 40, no. 278 (April 1767): 188–89, British Periodicals, https://www.proquest.com/britishperiodicals/docview/6194915/503E800D6D65491APQ; Ludvig Hektoen, "Experimental Measles," *Journal of Infectious Diseases* 2, no. 2 (1905): 238–55, https://academic.oup.com/jid/article/2/2/238/887192.

77 **subsequent reports of measles inoculation:** Hektoen, "Experimental Measles."

77 **"gathered courage to test the method":** Charles Herrman, "Immunization Against Measles," *Archives of Pediatrics* 32 (1915): 503–507.

77 **measles-like illness in monkeys:** John F. Anderson and Joseph Goldberger, "Recent Advances in Our Knowledge of Measles," *American Journal of Diseases of Children* 4, no. 1 (1912): 20–26, doi:10.1001/archpedi.1912.04100190023004.

78 **neither contracted measles:** Notably, a similar set of experiments using inoculation of young calves born to cows that had recovered from rinderpest was carried out by Geert

NOTES

Reinders in the Netherlands in 1774. Reviewed in Huygelen, "The Immunization of Cattle Against Rinderpest in Eighteenth-Century Europe."

79 **an unalloyed success:** Herrman, "Immunization Against Measles."

79 **"intentionally expose children to infection":** Herrman, "Immunization Against Measles," 505.

80 **a cautionary tale for mothers:** William Colby Rucker, "Measles Make Many Mothers Mourn," *Bridgeport Evening Farmer*, November 22, 1913, 2, https://chroniclingamerica.loc.gov/lccn/sn84022472/1913-11-22/ed-1/seq-2/.

80 **"every known measure for the prevention":** William Colby Rucker, *Measles*, supplement no. 1 to the Public Health Reports, January 24, 1913 [edition of June 1916] (Washington, DC: Government Printing Office, 1916), https://www.gutenberg.org/ebooks/19965.

81 **"The Etiology of Measles":** J. F. Enders, "The Etiology of Measles," in *Virus and Rickettsial Diseases, with Especial Consideration of Their Public Health Significance* (Cambridge: Harvard University Press, 1940), chap. 9, 237–67. There is a detailed (and petty) review of the symposium volume here: Thomas M. Rivers, *Science* 91, no. 2356 (1940): 192–94, https://www.science.org/doi/abs/10.1126/science.91.2356.192.

82 **Some trials seemed to be successful:** Enders, "The Etiology of Measles."

82 **protection against an epidemic microbial scourge:** David M. Oshinsky, *Polio: An American Story* (New York: Oxford University Press, 2006).

83 **cells from intestines:** T. H. Weller et al., "Studies on the Cultivation of Poliomyelitis Viruses in Tissue Culture. I. The Propagation of Poliomyelitis Viruses in Suspended Cell Cultures of Various Human Tissues," *Journal of Immunology* 69, no. 6 (1952): 645–71, PMID: 13022978; F. C. Robbins, T. H. Weller, and J. F. Enders, "Studies on the Cultivation of Poliomyelitis Viruses in Tissue Culture. II. The Propagation of the Poliomyelitis Viruses in Roller-Tube Cultures of Various Human Tissues," *Journal of Immunology* 69, no. 6 (1952): 673–94, PMID: 13022979.

83 **shared a Nobel Prize:** The Nobel Prize in Physiology or Medicine 1954, https://www.nobelprize.org/prizes/medicine/1954/summary/.

84 **found measles more interesting:** Samuel Lawrence Katz, American Academy of Pediatrics Oral History Project, https://downloads.aap.org/AAP/Gartner%20Pediatric%20History/Katz.pdf, 17.

84 **He accepted Enders's offer:** Douglas Martin, "Dr. Thomas C. Peebles, Who Identified Measles Virus, Dies at 89," *New York Times*, August 4, 2010, https://www.nytimes.com/2010/08/05/health/05peebles.html; Greer Williams, *Virus Hunters* (New York: Knopf, 1959).

84 **to pinpoint its cause:** Theodore H. Ingalls, Roswell Gallagher, and John F. Enders, "An Outbreak of Influenza A in a Boys' School," *New England Journal of Medicine* 235, no. 22 (1946): 786–88, doi: 10.1056/NEJM194611282352204, PMID: 20274906.

85 **"standing on the frontiers of science":** Williams, *Virus Hunters*, 383.

85 **Participation in the study:** John F. Enders and Thomas C. Peebles, "Propagation in Tissue Cultures of Cytopathogenic Agents from Patients with Measles," *Proceedings of the Society for Experimental Biology and Medicine* 86, no. 2 (1954): 277–86, doi: 10.3181/00379727-86-21073, PMID: 13177653.

NOTES

85 **"I said yes, of course":** James Glenday, "The Boy Behind the Measles Vaccine Grew Up to Be an Anti-Vaxxer. Now He Wants Everyone to Get Vaccinated," ABC News, May 11, 2019, https://www.abc.net.au/news/2019-05-12/meet-the-man-whose-illness-led-to-the-measles-vaccine/11100052.

86 **survive and reproduce in captivity:** Enders and Peebles, "Propagation in Tissue Cultures of Cytopathogenic Agents from Patients with Measles."

86 **Just like humans:** Thomas C. Peebles et al., "Behavior of Monkeys After Inoculation of Virus Derived from Patients with Measles and Propagated in Tissue Culture Together with Observations on Spontaneous Infections of These Animals by an Agent Exhibiting Similar Antigenic Properties," *Journal of Immunology* 78, no. 1 (1957): 63–74, PMID: 13406267.

86 **learned his lesson:** Samuel Lawrence Katz, American Academy of Pediatrics Oral History Project.

87 **made a B strain as well:** Samuel L. Katz, John F. Enders, and Ann Holloway, "The Development and Evaluation of an Attenuated Measles Virus Vaccine," *American Journal of Public Health and the Nation's Health* 52, suppl. 2 (1962): 5–10, doi: 10.2105/ajph.52.suppl_2.5, PMID: 14454407; John F. Enders et al., "Studies on an Attenuated Measles-Virus Vaccine. I. Development and Preparations of the Vaccine: Technics for Assay of Effects of Vaccination," *New England Journal of Medicine* 263 (1960): 153–59, doi: 10.1056/NEJM196007282630401, PMID: 13820246.

88 **Self-experimentation has a long history:** Lawrence K. Altman, *Who Goes First? The Story of Self-Experimentation in Medicine* (New York: Random House, 1987).

88 **"you make it, you take it":** Barbara Kuter, "Maurice Hilleman and the MMR Vaccine," presentation given at European Society of Clinical Microbiology and Infectious Diseases (ESCMID) Conference on the Impact of Vaccines on Public Health, Prague, Czechia, 3 April 2011.

89 **strange dichotomy:** Paul A. Offit, *Vaccinated: One Man's Quest to Defeat the World's Deadliest Diseases* (Washington, DC: Smithsonian Books, 2007), 24–25.

89 **Consent was often cursory:** Allen M. Hornblum, Judith L. Newman, and Gregory J. Dober, *Against Their Will: The Secret History of Medical Experimentation on Children in Cold War America* (New York: Palgrave Macmillan, 2013).

89 **no option for refusal:** Offit, *Vaccinated: One Man's Quest to Defeat the World's Deadliest Diseases*, 126.

90 **considered genetically inferior:** Marie E. Daly, "History of the Walter E. Fernald Development Center," https://www.city.waltham.ma.us/sites/g/files/vyhlif6861/f/file/file/fernald_center_history.pdf.

90 **notorious and well-documented history:** Hornblum, Newman, and Dober, *Against Their Will: The Secret History of Medical Experimentation on Children in Cold War America*.

90 **oatmeal laced with radioactive isotopes:** Advisory Committee on Human Radiation Experiments, "The Studies at the Fernald School," chapter 7 in Advisory Committee on Human Radiation Experiments—Final Report, no. 061-000-00-848-9 (Washington, DC: US Government Printing Office, 1995), https://bioethicsarchive.georgetown.edu/achre/final/chap7_5.html; G. S. Kurland et al., "Radioisotope Study of Thyroid Function

NOTES

in 21 Mongoloid Subjects, Including Observations in 7 Parents," *Journal of Clinical Endocrinology and Metabolism* 17, no. 4 (1957): 552–60, doi: 10.1210/jcem-17-4-552, PMID: 13406017; "Radioactive Oatmeal Suit Settled for $1.85 Million," *Washington Post*, January 1, 1998, https://www.washingtonpost.com/archive/politics/1998/01/01/radioactive-oatmeal-suit-settled-for-185-million/93894a5a-5844-4544-aca2-ffe4e52030b3; Hornblum, Newman, and Dober, *Against Their Will: The Secret History of Medical Experimentation on Children in Cold War America*.

91 **lively discussion followed:** Samuel L. Katz and John F. Enders, "Immunization of Children with a Live Attenuated Measles Virus," in S. S. Gellis et al., Transactions of the Society for Pediatric Research: Twenty-Ninth Annual Meeting. *American Journal of Diseases of Children* 98, no. 5 (1959): 553–681, doi:10.1001/archpedi.1959.02070020555002. Katz's remarks and the ensuing discussion are on pages 85–87 of the proceedings.

91 **more attenuation might be required:** John F. Enders, "Studies on the Virus of Measles," *Connecticut Medicine* 23, no. 9 (1959): 561–67.

92 **eight separate papers:** Measles-Vaccine Symposium, *New England Journal of Medicine* 263, no. 4 (1960): 202–203.

93 **"toxic as hell":** Offit, *Vaccinated: One Man's Quest to Defeat the World's Deadliest Diseases*, 48.

93 **low doses of gamma globulin:** Charles M. Reilly et al., "Living Attenuated Measles-Virus Vaccine in Early Infancy. Studies of the Role of Passive Antibody in Immunization," *New England Journal of Medicine* 265 (1961): 165–69, doi: 10.1056/NEJM196107272650403, PMID: 13740532. These studies are described in detail in Offit, *Vaccinated: One Man's Quest to Defeat the World's Deadliest Diseases*, 48–50.

94 **"protect children against a formidable":** John F. Kennedy, "Message from President John F. Kennedy to Dr. C. Henry Kempe," *American Journal of Diseases of Children* 103, no. 3 (1962): 213, doi:10.1001/archpedi.1962.02080020225001.

94 **provide a road map:** Luther L. Terry, "Opening Remarks," *American Journal of Diseases of Children* 103, no. 3 (1962): 217–18, doi:10.1001/archpedi.1962.02080020229003.

94 **"early eradication" of measles:** A. D. Langmuir, "Medical Importance of Measles," *American Journal of Diseases of Children* 103 (1962): 224–26, doi: 10.1001/archpedi.1962.02080020236005, PMID: 14462174.

95 **to evaluate three distinct products:** Roderick Murray, "Biologics Control of New Viral Vaccines: General Aspects," *American Journal of Diseases of Children* 103, no. 3 (1962): 434–37, doi:10.1001/archpedi.1962.02080020446058.

95 **U.S. licensure of two measles vaccines:** Louis Galambos with Jane Eliot Sewell, *Networks of Innovation: Vaccine Development at Merck, Sharp & Dohme, and Mulford, 1895–1995* (New York: Cambridge University Press, 1995), 96; Jan Hendriks and Stuart Blume, "Measles Vaccination Before the Measles-Mumps-Rubella Vaccine," *American Journal of Public Health* 103, no. 8 (2013): 1393–401, doi: 10.2105/AJPH.2012.301075, PMID: 23763422.

95 **emphasized the severity of measles:** "2 Measles Vaccines Licensed; U.S. Sees End of Disease in 1965," *New York Times*, March 22, 1963, https://www.nytimes.com/1963/03/22/archives/2-measles-vaccines-licensed-us-sees-end-of-disease-in-1965-us.html.

NOTES

95 **considerable coordination:** J. S. Hunter and B. E. Becker, "Telling the World About Measles; Case History in Government Information," *Public Health Reports* 73, no. 10 (1963): 893–96, PMID: 14062220.

96 **"quietly compelling drama":** Jack Gould, "'Taming of a Virus'; Compelling Documentary on Measles Offers Fascinating, Never Dry, Details," *New York Times*, March 25, 1963, https://www.nytimes.com/1963/03/25/archives/tv-taming-of-a-virus-compelling-documentary-on-measles-offers.html.

96 **no large-scale public vaccination drives:** "2 Drug Firms Licensed for Measles Vaccines: Live Strain Due on Market," *Los Angeles Times*, March 22, 1963, A5.

96 **private physician offices and clinics:** L. L. Terry, "The Status of Measles Vaccines. A Technical Report," *Journal of the National Medical Association* 55, no. 5 (1963): 453–55, PMID: 14049565.

96 **confused both parents and physicians:** H. Black, "Live Measles Vaccine Advocated by State—One Shot Does Job," *Boston Globe*, March 24, 1963, 8.

96 **greeted with jubilation:** Oshinsky, *Polio: An American Story*.

97 **funded the trial and purchased doses:** James Keith Colgrove, *State of Immunity: The Politics of Vaccination in Twentieth-Century America* (Berkeley: University of California Press, 2006), 117.

CHAPTER 7: VACCINES DON'T SAVE LIVES. VACCINATIONS SAVE LIVES.

99 **"the disease I love to hate":** Walter Orenstein, author interview, September 25, 2020; Walter A. Orenstein and Rafi Ahmed, "Simply Put: Vaccination Saves Lives," *Proceedings of the National Academy of Sciences of the USA* 114, no. 16 (2017): 4031–33, doi: 10.1073/pnas.1704507114, PMID: 28396427; Walter Orenstein, "Vaccines Don't Save Lives. Vaccinations Save Lives," *Human Vaccines & Immunotherapeutics* 15, no. 12 (2019): 2786–89, doi: 10.1080/21645515.2019.1682360, PMID: 31702427.

100 **In Kennedy's copy:** John F. Kennedy, "Annual Message to the Congress on the State of the Union, Speaking Copy, 11 January 1962," https://www.jfklibrary.org/asset-viewer/archives/JFKPOF/037/JFKPOF-037-003.

101 **funding and expertise to support state and local programs:** James Keith Colgrove, *State of Immunity: The Politics of Vaccination in Twentieth-Century America* (Berkeley: University of California Press, 2006), 146–47.

101 **leaving control in the hands of the states:** Alan R. Hinman, Walter A. Orenstein, and Lance Rodewald, "Financing Immunizations in the United States," *Clinical Infectious Diseases* 38, no. 10 (2004): 1440–46, doi: 10.1086/420748, PMID: 15156483.

101 **deliberately left the door open:** Elena Conis, *Vaccine Nation: America's Changing Relationship with Immunization* (Chicago: University of Chicago Press, 2015), 37–38.

101 **measles as an example:** House Hearing, 87th Congress—Intensive Immunization Programs, May 15 and 16, 1962, https://www.govinfo.gov/app/details/CHRG-87hhrg84426.

102 **signed into law:** "President Signs Foreign Aid Bill," *New York Times*, October 24, 1962, 18; Hinman, Orenstein, and Rodewald, "Financing Immunizations in the United States."

NOTES

102 **Some vaccine recommendations:** L. Reed Walton, Walter A. Orenstein, and Larry K. Pickering, "The History of the United States Advisory Committee on Immunization Practices (ACIP)," *Vaccine* 33, no. 3 (2015): 405–14, doi: 10.1016/j.vaccine.2014.09.043, PMID: 25446820.

102 **Advisory Committee on Immunization Practices:** Jean Clare Smith, Alan R. Hinman, and Larry K. Pickering, "History and Evolution of the Advisory Committee on Immunization Practices—United States, 1964–2014," *Morbidity and Mortality Weekly Report* 63, no. 42 (2014): 955–58, PMID: 25340913.

103 **delivered 5 million doses:** J. J. Goldman, "New Measles Vaccines Fail to Curb Incidence of Disease This Year," *Wall Street Journal*, June 25, 1964, 1.

103 **likely to be long and difficult:** Goldman, "New Measles Vaccines Fail to Curb Incidence of Disease This Year."

103 **coproduced a twenty-minute film:** Mission, "Measles: The Story of a Vaccine," https://collections.nlm.nih.gov/catalog/nlm:nlmuid-8901910A-vid.

103 **increasing the relative burden of disease:** Colgrove, *State of Immunity: The Politics of Vaccination in Twentieth-Century America*, 151.

104 **added measles to the list of included immunizations:** Conis, *Vaccine Nation: America's Changing Relationship with Immunization*, 48; 79 Statute 435, "An Act to extend and otherwise amend certain expiring provisions of the Public Health Service Act relating to community health services, and for other purposes," https://www.govinfo.gov/app/details/STATUTE-79/STATUTE-79-Pg435.

104 **Making measles vaccine eligible:** Alan R. Hinman, Walter A. Orenstein, and Mark J. Papania, "Evolution of Measles Elimination Strategies in the United States," *Journal of Infectious Diseases* 189, suppl. 1 (2004): S17–S22, doi: 10.1086/377694, PMID: 15106084.

104 **calling for the eradication of measles:** David J. Sencer, H. B. Dull, and A.D. Langmuir, "Epidemiologic Basis for Eradication of Measles in 1967," *Public Health Reports* 82, no. 3 (1967): 253–56, PMID: 4960501; David J. Sencer, "A Program to Eradicate Measles," *American Journal of Public Health and the Nation's Health* 57, no. 5 (1967): 729–30, doi: 10.2105/ajph.57.5.729-b, PMID: 18018172.

104 **four overlapping strategies:** Sencer, Dull, and Langmuir, "Epidemiologic Basis for Eradication of Measles in 1967."

104 **"should no longer be tolerated":** Sencer, "A Program to Eradicate Measles."

104 **"Measles, so familiar in our youth":** Lyndon B. Johnson, "Proclamation 3806—Child Health Day, 1967," Gerhard Peters and John T. Woolley, the American Presidency Project, https://www.presidency.ucsb.edu/node/306287; https://www.govinfo.gov/content/pkg/STATUTE-81/pdf/STATUTE-81-Pg1128.pdf, 1129.

105 **remaining a threat:** W. R. Dowdle, "The Principles of Disease Elimination and Eradication," *Bulletin of the World Health Organization* 76, suppl. 2 (1998): 22–25, PMID: 10063669.

105 **not recommended until 1968:** Jan Hendriks and Stuart Blume, "Measles Vaccination Before the Measles-Mumps-Rubella Vaccine," *American Journal of Public Health* 103, no. 8 (2013): 1393–401, doi: 10.2105/AJPH.2012.301075, PMID: 23763422.

NOTES

105 **"a horrendous disease":** William H. Foege, *The Fears of the Rich, the Needs of the Poor: My Years at the CDC* (Baltimore: Johns Hopkins University Press, 2018), 75.

106 **local resurgences of measles:** Foege, *The Fears of the Rich, the Needs of the Poor: My Years at the CDC*, 77.

106 **weeklong series of *Peanuts* cartoons:** Colgrove, *State of Immunity: The Politics of Vaccination in Twentieth-Century America*, 159–60. The cartoon series ran from January 2 to January 7, 1967, https://www.gocomics.com/peanuts/1967/01/02.

106 **"Measles Have Just About Had It":** Harold M. Schmeck Jr., "Measles Have Just About Had It," *New York Times*, March 26, 1967, https://www.nytimes.com/1967/03/26/archives/measles-have-just-about-had-it.html.

106 **reported measles cases dropped:** Hinman, Orenstein, and Papania, "Evolution of Measles Elimination Strategies in the United States."

106 **would likely take several more years:** "Medical News," *Journal of the American Medical Association* 209, no. 2 (1969): 191–200, doi:10.1001/jama.1969.03160150005003.

106 **families in the "hard-core ghetto":** "Medical News," *Journal of the American Medical Association* 209, no. 2 (1969): 191–200, doi:10.1001/jama.1969.03160150005003.

107 **elimination of federal funding:** J. L. Conrad, Robert Wallace, and John J. Witte, "The Epidemiologic Rationale for the Failure to Eradicate Measles in the United States," *American Journal of Public Health* 61, no. 11 (1971): 2304–10, doi: 10.2105/ajph.61.11.2304, PMID: 5112980.

107 **unfunded or underfunded:** Robert J. Bazell, "Health Programs: Slum Children Suffer Because of Low Funding," *Science* 172, no. 3986 (1971): 921–25, doi: 10.1126/science.172.3986.921, PMID: 5103301.

107 **sometimes called "German measles":** "German measles" terminology is included for historical interest. Using place names to designate infectious agents is (fortunately) falling out of fashion, in part because of the xenophobia discussed in later chapters.

108 **Prioritization of the emerging vaccines for rubella:** A. R. Hinman, A. D. Brandling-Bennett, and P. I. Nieburg, "The Opportunity and Obligation to Eliminate Measles from the United States," *Journal of the American Medical Association* 242, no. 11 (1979): 1157–62, PMID: 470069.

108 **interruption of indigenous measles transmission:** Hinman, Brandling-Bennett, and Nieburg, "The Opportunity and Obligation to Eliminate Measles from the United States."

109 **the same local businesses, churches, and events:** P. J. Landrigan, "Epidemic Measles in a Divided City," *Journal of the American Medical Association* 221, no. 6 (1972): 567–70, PMID: 5068077; CDC, "Measles—Texarkana, Texas, and Arkansas," *Morbidity and Mortality Weekly Report* 20, no. 6 (1971): 46-47, https://stacks.cdc.gov/view/cdc/1735; CDC, Measles Surveillance Report 1971, https://www.google.com/books/edition/_/4eRPymp0tEAC.

110 **alter the life expectancy:** Jonathan Metzl, *Dying of Whiteness: How the Politics of Racial Resentment Is Killing America's Heartland* (New York: Basic Books, 2019).

NOTES

111 **widening divide in health care access:** Kim Krisberg. "In Two-State Texarkana, a Widening Divide in Health Care Access," *Public Health Watch*, April 10, 2023, https://public healthwatch.org/2023/04/10/medicaid-expansion-texarkana-texas-arkansas-health.

CHAPTER 8: IMPERFECT TOOLS

114 **"The whole mess stinks":** Robert Kistler, "Measles Epidemic a Product of Neglect," *Los Angeles Times*, April 1, 1977, A30.

114 **Additional vaccines and supplies arrived:** Robert Kistler, "Emergency Vaccine Arrives for Anti-Measles Drive: County Fight Against Measles," *Los Angeles Times*, April 2, 1977, A1.

114 **"last-minute" measles vaccination:** Bill Stall, "35,000 Students Get Last-Minute Measles Shots," *Los Angeles Times*, May 1, 1977, 1.

114 **barred about 40,000 students:** Jack McCurdy and H. Nelson, "Schools Bar Thousands Lacking Measles Shots," *Los Angeles Times*, May 3, 1977, 15.

114 **remained on the exclusion list:** McCurdy and Nelson, "Schools Bar Thousands Lacking Measles Shots."

114 **overall toll was substantial:** J. D. Cherry, "Measles: A Plague That Was Preventable," *Los Angeles Times*, May 1, 1977, I1; McCurdy and Nelson, "Schools Bar Thousands Lacking Measles Shots."

115 **lack of awareness:** Stall B. "35,000 Students Get Last-Minute Measles Shots," *Los Angeles Times*, May 1, 1977, 1; H. Nelson and R. Kistler, "Schools to Bar Pupils Lacking Measles Shots: County Crackdown on Measles," *Los Angeles Times*, April 1, 1977, 1.

115 **underfunded federal programs:** Patrick Michael Vivier, "National Policies for Childhood Immunization in the United States: An Historical Perspective," thesis, Johns Hopkins University, 1996, 164–65. https://www.proquest.com/docview/304306792?pq-origsite=gscholar&fromopenview=true.

115 **support for the idea of national health insurance:** "Carter Proposes U.S. Health Plan," *New York Times*, April 17, 1976, https://www.nytimes.com/1976/04/17/archives/carter-proposes-us-health-plan-says-he-favors-mandatory-insurance.html.

116 **not much would be required:** AP, "Bumpers' Wife Is Reluctant to Leave Charleston Home," *Shreveport Times*, November 22, 1970, 18-A.

116 **"Every Child by '74" campaign:** Testimony of Senator Bumpers, Supplemental Appropriations for Fiscal Year 1977—Hearings Before Subcommittees of the Committee on Appropriations, House of Representatives, 95th Congress, First Session, Part 2, https://www.google.com/books/edition/Supplemental_Appropriations_for_Fiscal_Y/o81Ozj8NfXkC?hl=en&gbpv=0, 574; Dale Bumpers, *The Best Lawyer in a One-Lawyer Town: A Memoir* (New York: Random House, 2003), 223–24.

116 **Bumpers's Arkansas program:** Testimony of Senator Bumpers, Supplemental Appropriations for Fiscal Year 1977—Hearings Before Subcommittees of the Committee on Appropriations, House of Representatives, 95th Congress, First Session, Part 2, 574.

116 **immunizing more than 300,000 Arkansas children:** Bumpers, *The Best Lawyer in a One-Lawyer Town: A Memoir*, 223–24.

NOTES

117 **Califano and Bumpers met:** Joseph A. Califano Jr., *Governing America: An Insider's Report from the White House and the Cabinet* (New York: Simon & Schuster, 1981), 179.

117 **one of the most abysmal failures:** Testimony of Senator Bumpers. Supplemental Appropriations for Fiscal Year 1977—Hearings Before Subcommittees of the Committee on Appropriations, House of Representatives, 95th Congress, First Session, Part 2, 571.

117 **substantial intellectual debt:** "Cohn v. Califano Sets '79 Deadline for Immunizing Children," *Washington Post*, April 7, 1977; Califano, *Governing America*, 179.

118 **cornerstone of the strategy:** Alan R. Hinman, "The New U.S. Initiative in Childhood Immunization," *Bulletin of the Pan American Health Organization* 13, no. 2 (1979): 169–76, PMID: 454866; A. R. Hinman, A. D. Brandling-Bennett, and P. I. Nieburg, "The Opportunity and Obligation to Eliminate Measles from the United States," *Journal of the American Medical Association* 242, no. 11 (1979): 1157–62, PMID: 470069.

118 **their own materials:** Meredith A. Hickson, "National Level Immunization Promotion Campaigns," *1982 17th Immunization Conference Proceedings*, May 18–19, 1982, Atlanta, Georgia, https://www.google.com/books/edition/Immunization_Conference_Proceedings/YjMKvB7krqYC?hl=en, 107–108.

118 **get messages to children and their parents:** Vivier, "National Policies for Childhood Immunization in the United States: An Historical Perspective," 170; Elena Conis, *Vaccine Nation: America's Changing Relationship with Immunization* (Chicago: University of Chicago Press, 2015), 97–98.

119 **stayed concerningly low:** Walter A. Orenstein, "The Role of Measles Elimination in Development of a National Immunization Program," *Pediatric Infectious Disease Journal* 25, no. 12 (2006): 1093–101, doi: 10.1097/01.inf.0000246840.13477.28, PMID: 17133153.

119 **"familiar-sounding goal":** Elena Conis, "Measles and the Modern History of Vaccination," *Public Health Reports* 134, no. 2 (2019): 118–25, doi: 10.1177/0033354919826558, PMID: 30763141.

119 **"elimination of indigenous measles":** Hinman, Brandling-Bennett, and Nieburg, "The Opportunity and Obligation to Eliminate Measles from the United States."

120 **laws mandating immunizations:** Alan R. Hinman et al., "Progress in Measles Elimination," *Journal of the American Medical Association* 247, no. 11 (1982): 1592–95, PMID: 7062463.

122 **misalignment of incentives:** Paul Starr, *The Social Transformation of American Medicine: The Rise of a Sovereign Profession and the Making of a Vast Industry* (New York: Basic Books, 1982), 379.

122 **unmasking of the Tuskegee experiments:** Jean Heller, "Syphilis Victims in U.S. Study Went Untreated for 40 Years," *New York Times*, July 26, 1972, 1, https://www.nytimes.com/1972/07/26/archives/syphilis-victims-in-us-study-went-untreated-for-40-years-syphilis.html; Susan Reverby, *Examining Tuskegee: The Infamous Syphilis Study and Its Legacy* (Chapel Hill: University of North Carolina Press, 2009); David S. Jones, Christine Grady, and Susan E. Lederer, "'Ethics and Clinical Research'—The 50th Anniversary of Beecher's Bombshell," *New England Journal of Medicine* 374, no. 24 (2016): 2393–98, doi: 10.1056/NEJMms1603756, PMID: 27305197.

NOTES

123 **secretive national program of intentional infection:** David J. Rothman and Sheila M. Rothman, *The Willowbrook Wars* (New York: Harper & Row, 1984); Sydney A. Halpern, *Dangerous Medicine: The Story Behind Human Experiments with Hepatitis* (New Haven: Yale University Press, 2021).

123 **eager to use his established population:** Samuel Lawrence Katz, American Academy of Pediatrics Oral History Project, https://downloads.aap.org/AAP/Gartner%20Pediatric%20History/Katz.pdf.

123 **deliberate infection of children:** Allen M. Hornblum, Judith L. Newman, and Gregory J. Dober, *Against Their Will: The Secret History of Medical Experimentation on Children in Cold War America* (New York City: Palgrave Macmillan, 2013), 99; Saul Krugman and Joan P. Giles, "Viral Hepatitis, Type B (MS-2-Strain). Further Observations on Natural History and Prevention," *New England Journal of Medicine* 288, no. 15 (1973): 755–60, doi: 10.1056/NEJM197304122881503, PMID: 4688714; Rothman and Rothman, *The Willowbrook Wars*; Halpern, *Dangerous Medicine: The Story Behind Human Experiments with Hepatitis*.

123 **highest accolades in academic medicine:** J. Dancis, "Presentation of the Howland Award to Dr. Saul Krugman," *Pediatric Research* 15, no. 10 (1981): 1323–27, doi: 10.1203/00006450-198110000-00002, PMID: 7029440; Harold M. Schmeck Jr., "Five U.S. Scientists Win 1983 Lasker Award," *New York Times*, November 17, 1983, https://www.nytimes.com/1983/11/17/us/five-us-scientists-win-1983-lasker-award.html.

124 **more than 150 people protested:** Harold M. Schmeck Jr., "Researcher, Target of a Protest, Is Lauded at Physicians' Parley," *New York Times*, April 18, 1972, https://www.nytimes.com/1972/04/18/archives/researcher-target-of-a-protest-is-lauded-at-physicians-parley.html; flyer from the protests here: "'Doctor' gives kids hepatitis. Stop racist Krugman," American University Archive Patrick Frazier Political and Social Movements Collection, https://dra.american.edu/islandora/object/auislandora%3A70871?solr_nav%5Bid%5D=79a9d45cbf84052b74d6&solr_nav%5Bpage%5D=2&solr_nav%5Boffset%5D=2.

124 **chronicled in detail:** Paul A. Offit, *The Cutter Incident: How America's First Polio Vaccine Led to the Growing Vaccine Crisis* (New Haven: Yale University Press, 2005).

125 **combination vaccines that protected:** Both DPT and DTP are used as abbreviations for the combination vaccine that includes antigens from diphtheria, tetanus, and whole-cell pertussis. I have generally used DTP throughout this work, as that abbreviation and its derivative (DTaP, which contains the acellular pertussis preparation) are more commonly used today.

125 **significant and underappreciated dangers:** Paul A. Offit, *Deadly Choices: How the Anti-Vaccine Movement Threatens Us All* (New York: Basic Books, 2011).

126 **published in a mainstream medical journal:** M. Kulenkampff, J. S. Schwartzman, and J. Wilson, "Neurological Complications of Pertussis Inoculation," *Archives of Disease in Childhood* 49, no. 1 (1974): 46–49, doi: 10.1136/adc.49.1.46, PMID: 4818092.

126 **a program called *DPT: Vaccine Roulette*:** Donna Hilts, "TV Report on Vaccine Stirs Bitter Controversy," *Washington Post*, April 28, 1982, https://www.washingtonpost.com

NOTES

/archive/local/1982/04/28/tv-report-on-vaccine-stirs-bitter-controversy/80d1fc8a-1012-4732-a517-7976c86ab52d/.

126 **the birth of "the modern American anti-vaccine movement":** Offit, *Deadly Choices: How the Anti-Vaccine Movement Threatens Us All*, 2.

126 **provides less durable pertussis prevention:** Stanley A. Plotkin, "The Pertussis Problem," *Clinical Infectious Diseases* 58, no. 6 (2014): 830–33, doi: 10.1093/cid/cit934, PMID: 24363332.

126 **catalyzing the formation of an anti-vaccine organization:** James Keith Colgrove, *State of Immunity: The Politics of Vaccination in Twentieth-Century America* (Berkeley: University of California Press, 2006), 211.

126 **rebroadcast widely:** Hilts, "TV Report on Vaccine Stirs Bitter Controversy."

127 **National Childhood Vaccine Injury Act:** H.R. 5184—National Childhood Vaccine Injury Act of 1986, https://www.congress.gov/bill/99th-congress/house-bill/5184.

127 **crucial to the survival of the U.S. immunization program:** M. H. Smith, "National Childhood Vaccine Injury Compensation Act," *Pediatrics* 82, no. 2 (1988): 264–69, PMID: 3399300; Katherine M. Cook and Geoffrey Evans, "The National Vaccine Injury Compensation Program," *Pediatrics* 127, suppl. 1 (2011): S74–S77, doi: 10.1542/peds.2010-1722K, PMID: 21502255; H. Cody Meissner, Narayan Nair, and Stanley A. Plotkin, "The National Vaccine Injury Compensation Program: Striking a Balance Between Individual Rights and Community Benefit," *JAMA* 321, no. 4 (2019): 343–44, doi: 10.1001/jama.2018.20421, PMID: 30608524.

127 **Vaccine Adverse Event Reporting System:** Centers for Disease Control and Prevention (CDC), "Vaccine Adverse Event Reporting System—United States," *Morbidity and Mortality Weekly Report* 39, no. 41 (1990): 730–33, PMID: 2120567; Frederick Varricchio et al., "Understanding Vaccine Safety Information from the Vaccine Adverse Event Reporting System," *Pediatric Infectious Disease Journal* 23, no. 4 (2004): 287–94, doi: 10.1097/00006454-200404000-00002, PMID: 15071280; Tom T. Shimabukuro et al., "Safety Monitoring in the Vaccine Adverse Event Reporting System (VAERS)," *Vaccine* 33, no. 36 (2015): 4398–405, doi: 10.1016/j.vaccine.2015.07.035, PMID: 26209838.

128 **sources of confusion:** Shimabukuro et al., "Safety Monitoring in the Vaccine Adverse Event Reporting System (VAERS)"; H. Cody Meissner, "Understanding Vaccine Safety and the Roles of the FDA and the CDC," *New England Journal of Medicine* 386, no. 17 (2022): 1638–45, doi: 10.1056/NEJMra2200583, PMID: 35476652.

128 **"VAERS dumpster dives":** David Gorski, "Dumpster Diving in the VAERS Database to Find More COVID-19 Vaccine-Associated Myocarditis in Children," *Science-Based Medicine*, September 13, 2021, https://sciencebasedmedicine.org/dumpster-diving-in-vaers-doctors-fall-into-the-same-trap-as-antivaxxers/.

129 **health departments had to make difficult choices:** Conis, *Vaccine Nation: America's Changing Relationship with Immunization*, 164.

129 **skewed both younger and older:** Lauri E. Markowitz et al., "Patterns of Transmission in Measles Outbreaks in the United States, 1985–1986," *New England Journal of Medicine* 320, no. 2 (1989): 75–81, doi: 10.1056/NEJM198901123200202, PMID: 2911293.

NOTES

129 **generally saw one or two measles cases:** A. Hayes, "Measles Warning Released," *Daily Kent Stater*, October 5, 1988, 1.

130 **revisited the Kent State measles outbreak:** Tara C. Smith, "When Measles Came to Class: A Look Back at the 1989 Kent State University Measles Epidemic," *Microbes and Infection* 24, no. 2 (2022): 104885, doi: 10.1016/j.micinf.2021.104885, PMID: 34536576.

130 **considerably lower risk of measles outbreaks:** Andrew L. Baughman et al., "The Impact of College Prematriculation Immunization Requirements on Risk for Measles Outbreaks," *Journal of the American Medical Association* 272, no. 14 (1994): 1127–32; doi:10.1001/jama.1994.03520140057038.

131 **a different kind of cautionary tale:** Bradley S. Hersh et al., "A Measles Outbreak at a College with a Prematriculation Immunization Requirement," *American Journal of Public Health* 81, no. 3 (1991): 360–64, doi: 10.2105/ajph.81.3.360, PMID: 1994745.

133 **an office in the Sears Tower:** T. Randall, "Measles Origin Still a Mystery," *Chicago Tribune*, August 6, 1989, 29.

133 **confirmed 262 cases:** "Measles Outbreak Hits 262," *Chicago Tribune*, July 10, 1989, N4.

133 **over a thousand:** Centers for Disease Control and Prevention (CDC), "Measles Outbreak—Chicago, 1989," *Morbidity and Mortality Weekly Report* 38, no. 34 (1989): 591–92, PMID: 2503702.

133 **Intensive public health measures:** M. L. Lindegren et al., "Measles Vaccination in Pediatric Emergency Departments During a Measles Outbreak," *Journal of the American Medical Association* 270, no. 18 (1993): 2185–89, PMID: 8411600; Jorge Casuso, "City Goes Door to Door in Fight Against Measles," *Chicago Tribune*, August 2, 1989, 1.

134 **more than doubled again:** Centers for Disease Control and Prevention (CDC), "Update: Measles Outbreak—Chicago, 1989," *Morbidity and Mortality Weekly Report* 39, no. 19 (1990): 317–9, 325–6, PMID: 2110291.

134 **more than ten times the rate:** CDC, "Update: Measles Outbreak—Chicago, 1989."

134 **classified by public health officials:** Race and ethnicity categories are presented here as they exist in the reports from the time. The terminology and methods for determining who falls into which category are far from ideal, but these are the only available data about an important issue of disparities.

135 **seven to nine times the rate of white children:** "The Measles Epidemic. The Problems, Barriers, and Recommendations, The National Vaccine Advisory Committee," *Journal of the American Medical Association* 266, no. 11 (1991): 1547–52, PMID: 1880887; Centers for Disease Control and Prevention (CDC), "Measles—United States, 1990," *Morbidity and Mortality Weekly Report* 40, no. 22 (1991): 369–72. PMID, 2034203.

135 **underlying causes of the measles epidemic:** Summarized in Nationwide Measles Epidemic Hearing Before the Subcommittee on Health and the Environment of the Committee on Energy and Commerce, House of Representatives, 102nd Congress, First Session, March 11, 1991, volume 4. (The formal reference is "The Measles Epidemic. The Problems, Barriers, and Recommendations, The National Vaccine Advisory Committee.") Also Philip J. Hilts, "Panel Ties Measles Epidemic to Breakdown in Health System," *New York*

NOTES

Times, January 9, 1991, https://www.nytimes.com/1991/01/09/us/panel-ties-measles-epidemic-to-breakdown-in-health-system.html.

135 **"a warning flag of problems with our system of primary health care"**: "The Measles Epidemic. The Problems, Barriers, and Recommendations, The National Vaccine Advisory Committee"; Walter A. Orenstein, "The Role of Measles Elimination in Development of a National Immunization Program," *Pediatric Infectious Disease Journal* 25, no. 12 (2006): 1093–101, doi: 10.1097/01.inf.0000246840.13477.28, PMID: 17133153.

136 **placed health care policy at the center:** Conis, *Vaccine Nation: America's Changing Relationship with Immunization,* chap. 7; Colgrove, *State of Immunity: The Politics of Vaccination in Twentieth-Century America,* 222.

138 **Immunization rates among preschool-aged children soared:** Alan R. Hinman, Walter A. Orenstein, and Lance Rodewald, "Financing Immunizations in the United States," *Clinical Infectious Diseases* 38, no. 10 (2004): 1440–46, doi: 10.1086/420748, PMID: 15156483; J. M. Santoli et al., "Vaccines for Children Program, United States, 1997," *Pediatrics* 104, no. 2 (1999): e15, doi: 10.1542/peds.104.2.e15, PMID: 10429133.

138 **its first twenty years:** Cynthia G. Whitney et al., "Benefits from Immunization During the Vaccines for Children Program Era—United States, 1994–2013," *Morbidity and Mortality Weekly Report* 63, no. 16 (2014): 352–5, PMID: 24759657.

138 **disparities in immunization rates:** Brendan Walsh, Edel Doherty, and Ciaran O'Neill, "Since the Start of the Vaccines for Children Program, Uptake Has Increased, and Most Disparities Have Decreased," *Health Affairs* (Millwood), 35, no. 2 (2016): 356–64, doi: 10.1377/hlthaff.2015.1019, PMID: 26858392; Allison T. Walker, Philip J. Smith, and Maureen Kolasa, "Reduction of Racial/Ethnic Disparities in Vaccination Coverage, 1995–2011," *Morbidity and Mortality Weekly Report Suppl.* 63, no. 1 (2014): 7–12, PMID: 24743661; Sonja S. Hutchins, Ruth Jiles, and Roger Bernier, "Elimination of Measles and of Disparities in Measles Childhood Vaccine Coverage Among Racial and Ethnic Minority Populations in the United States," *Journal of Infectious Diseases* 189, suppl. 1 (2004): S146–S152, doi: 10.1086/379651, PMID: 15106103.

138 **no single strain of measles virus:** Paul A. Rota et al., "Genetic Analysis of Measles Viruses Isolated in the United States Between 1989 and 2001: Absence of an Endemic Genotype Since 1994," *Journal of Infectious Diseases* 189, suppl. 1 (2004): S160–S164, doi: 10.1086/374607, PMID: 15106105.

138 **CDC panel certified:** Walter A. Orenstein, Mark J. Papania, and Melinda E. Wharton, "Measles Elimination in the United States," *Journal of Infectious Diseases* 189, suppl. 1 (2004): S1–S3, doi: 10.1086/377693, PMID: 15106120; Samuel L. Katz and Alan R. Hinman, "Summary and Conclusions: Measles Elimination Meeting, 16–17 March 2000," *Journal of Infectious Diseases* 189, suppl. 1 (2004): S43–S47, doi: 10.1086/377696, PMID: 15106088.

139 **increasingly ambitious goals:** Anna Maria Henao-Restrepo et al., "Experience in Global Measles Control, 1990–2001," *Journal of Infectious Diseases* 187, suppl. 1 (2003): S15–S21, doi: 10.1086/368273, PMID: 12721887.

NOTES

139 **ten of those fourteen were in Africa:** "Measles: Progress Towards Global Control and Regional Elimination, 1990–1998," *Weekly Epidemiological Record* 73, no. 50 (1998): 389–94, PMID: 9868907; Centers for Disease Control and Prevention (CDC), "Global Measles Control and Regional Elimination, 1998–1999," *Morbidity and Mortality Weekly Report* 48, no. 49 (1999): 1124–30, PMID: 10634347.

140 **had already interrupted endemic measles transmission:** Henao-Restrepo et al., "Experience in Global Measles Control, 1990–2001."

140 **measles might be eradicated globally:** Donald R. Hopkins et al., "The Case for Global Measles Eradication," *Lancet* 319, no. 8286 (1982): 1396–98, doi: 10.1016/s0140-6736(82)92510-7, PMID: 6123687; William H. Foege, "The Global Elimination of Measles," *Public Health Reports* 97, no. 5 (1982): 402–405, PMID: 7122818.

140 **"worthy of our best endeavors":** Foege, "The Global Elimination of Measles."

141 **"make measles vaccine a tugboat":** Global Eradication of Polio and Measles: Hearing Before a Subcommittee of the Committee on Appropriations, United States Senate, 105th Congress, Second Session, September 23, 1998, https://www.google.com/books/edition/Global_Eradication_of_Polio_and_Measles/YR95i7dVHPsC?hl=en&gbpv=0.

CHAPTER 9: AMNESIA

144 **a state called immune amnesia:** Michael J. Mina et al., "Long-Term Measles-Induced Immunomodulation Increases Overall Childhood Infectious Disease Mortality," *Science* 348, no. 6235 (2015): 694–99, doi: 10.1126/science.aaa3662, PMID: 25954009; Michael J. Mina et al., "Measles Virus Infection Diminishes Preexisting Antibodies That Offer Protection from Other Pathogens," *Science* 366, no. 6465 (2019): 599–606, doi: 10.1126/science.aay6485, PMID: 31672891; Brigitta M. Laksono et al., "Measles Pathogenesis, Immune Suppression and Animal Models," *Current Opinion in Virology* 41 (2020): 31–37, doi: 10.1016/j.coviro.2020.03.002, PMID: 32339942.

144 **mechanisms behind immune amnesia:** Rory D. de Vries et al., "Measles Immune Suppression: Lessons from the Macaque Model," *PLoS Pathogens* 8, no. 8 (2012): e1002885, doi: 10.1371/journal.ppat.1002885, PMID: 22952446; Brigitta M. Laksono et al., "Studies into the Mechanism of Measles-Associated Immune Suppression During a Measles Outbreak in the Netherlands," *Nature Communications* 9, no. 1 (2018): 4944, doi: 10.1038/s41467-018-07515-0, PMID: 30470742; Mina et al., "Measles Virus Infection Diminishes Preexisting Antibodies That Offer Protection from Other Pathogens."

145 **unexpectedly had a negative result:** Stanford T. Shulman, "Clemens von Pirquet: A Remarkable Life and Career," *Journal of the Pediatric Infectious Diseases Society* 6, no. 4 (2017): 376–79, doi: 10.1093/jpids/piw063, PMID: 27794078; Clemens von Pirquet, "Das Verhalten der kutanen Tuberkulinreaktion während der Masern," *Deutsche Medizinische Wochenschrift* 34, no. 30 (1908): 1297–300, https://doi.org/10.1055/s-0028-1135624.

145 **period of increased vulnerability:** Mina et al., "Long-Term Measles-Induced Immunomodulation Increases Overall Childhood Infectious Disease Mortality"; Kartini Gadroen et al., "Impact and Longevity of Measles-Associated Immune Suppression: A

NOTES

Matched Cohort Study Using Data from the THIN General Practice Database in the UK," *BMJ Open* 8, no. 11 (2018): e021465, doi: 10.1136/bmjopen-2017-021465, PMID: 30413497.

145 **less likely to die not just from measles:** Julian P. T. Higgins et al., "Association of BCG, DTP, and Measles Containing Vaccines with Childhood Mortality: Systematic Review," *BMJ* 355 (2016): i5170, doi: 10.1136/bmj.i5170, PMID: 27737834.

146 **The committee's four answers:** "The Measles Epidemic. The Problems, Barriers, and Recommendations, The National Vaccine Advisory Committee," *Journal of the American Medical Association* 266, no. 11 (1991): 1547–52, PMID: 1880887.

147 **"people who do not object":** Problems of Vaccine Supply and Delivery—Hearing Before a Subcommittee of the Committee on Appropriations, United States Senate, 102nd Congress, First Session, Special Hearing 1991, https://www.google.com/books/edition/Problems_of_Vaccine_Supply_and_Delivery/j9gOW7ktTvgC?hl=en, 45.

147 **vaccine refusal did play a major role:** D. V. Rodgers et al., "High Attack Rates and Case Fatality During a Measles Outbreak in Groups with Religious Exemption to Vaccination," *Pediatric Infectious Disease Journal* 12, no. 4 (1993): 288–92, doi: 10.1097/00006454-199304000-00006, PMID: 8483622.

148 **first deaths among the children:** Paul A. Offit, *Bad Faith: When Religious Belief Undermines Modern Medicine* (New York: Basic Books, 2015).

148 **an unprecedented action:** Rodgers et al., "High Attack Rates and Case Fatality During a Measles Outbreak in Groups with Religious Exemption to Vaccination."

149 **linked Tuskegee's legacy to modern health disparities:** Jeff Stryker, "Tuskegee's Long Arm Still Touches a Nerve," *New York Times*, April 13, 1997, https://www.nytimes.com/1997/04/13/weekinreview/tuskegee-s-long-arm-still-touches-a-nerve.html; Donna St. George, "The Tuskegee Study's Legacy," *Philadelphia Inquirer*, August 15, 1993, A1.

149 **Direct comparisons:** T. Jones, "Are Black Kids 'Guinea Pigs'? Community Anger Intensifies After Judge Refuses to Halt Inoculations," *Tri-State Defender* (Memphis, TN), May 15, 1996, 1A.

149 **had not been approved for use:** Marlene Cimons, "CDC Says It Erred in Measles Study," *Los Angeles Times*, June 17, 1996, A11, https://www.washingtonpost.com/archive/politics/1996/06/17/us-measles-experiment-failed-to-disclose-risk/6a4dd6ce-7add-4daa-8c5e-6e5fc343b996/.

150 **came up repeatedly in the press:** B. Glenn, "Bad Blood Once Again," *St. Petersburg Times*, July 21, 1996, 5D; Abigail Trafford, "The Ghost of Tuskegee," *Washington Post*, May 6, 1997, A19, https://www.washingtonpost.com/archive/opinions/1997/05/06/the-ghost-of-tuskegee/51f033d2-bfd3-4431-a054-114c5d122c93/.

150 **Saul Krugman had died:** Wolfgang Saxon, "Saul Krugman, 84; Led Fight to Vanquish Childhood Diseases," *New York Times*, October 28, 1995, https://www.nytimes.com/1995/10/28/nyregion/saul-krugman-84-led-fight-to-vanquish-childhood-diseases.html.

150 **drawing a throughline:** David J. Rothman, "Government Guinea Pigs," *New York Times*, January 9, 1994, https://www.nytimes.com/1994/01/09/opinion/government-guinea-pigs.html; Sheryl Gay Stolberg, "When People Experiment on People," *Los Angeles Times*, January

NOTES

16, 1994, A1; Lisa M. Krieger, "Promise to 'Do No Harm' Often Abandoned," *San Francisco Examiner*, December 29, 1993, 12; David L. Kirp, "Blood, Sweat, and Tears: The Tuskegee Experiment and the Era of AIDS," *Tikkun* 10, no. 3 (1995): 50–54, PMID: 11657135.

150 **predictable spike in measles cases:** M. Ramsay, "The Epidemiology of Measles in England and Wales: Rationale for the 1994 National Vaccination Campaign," *Communicable Disease Report—CDR Review* 4, no. 12 (1994): R141–R146, PMID: 7529089.

150 **blamed MMR vaccination:** Seth Mnookin, *The Panic Virus: A True Story of Medicine, Science, and Fear* (New York: Simon & Schuster, 2011).

150 **published a report:** Andrew J. Wakefield et al., "Ileal-Lymphoid-Nodular Hyperplasia, Non-Specific Colitis, and Pervasive Developmental Disorder in Children," *Lancet* 351, no. 9103 (1998): 637–41, doi: 10.1016/s0140-6736(97)11096-0; retraction in *Lancet* 375, no. 9713 (2010): 445, PMID: 9500320.

151 **"peel the rancid onion":** Brian Deer, *The Doctor Who Fooled the World: Science, Deception, and the War on Vaccines* (Baltimore: Johns Hopkins University Press, 2020), 86.

151 **a multitude of subsequent studies:** Courtney Gidengil et al., "Safety of Vaccines Used for Routine Immunization in the United States: An Updated Systematic Review and Meta-Analysis," *Vaccine* 39, no. 28 (2021): 3696–716, doi: 10.1016/j.vaccine.2021.03.079, PMID: 34049735.

151 **"spectacularly wrong":** Paul A. Offit, "Junk Science Isn't a Victimless Crime," *Wall Street Journal*, January 11, 2011, https://www.wsj.com/articles/SB10001424052748703779704576073744290909186.

151 **plumbs the depths of Wakefield's deceptions:** Deer, *The Doctor Who Fooled the World: Science, Deception, and the War on Vaccines*, 86.

152 **longevity of Wakefield's message:** Heidi Larson, *Stuck: How Vaccine Rumors Start—and Why They Don't Go Away* (New York: Oxford University Press, 2020).

152 **continued to deny any wrongdoing:** Susan Dominus, "The Crash and Burn of an Autism Guru," *New York Times*, April 20, 2011, https://www.nytimes.com/2011/04/24/magazine/mag-24Autism-t.html.

152 **"He just knows how to speak":** Walter Orenstein, author interview, September 25, 2020.

152 **"I don't think that measles was as singled out":** Alan Hinman, author interview, October 27, 2020.

153 **testified at hearings:** Hearing Before the Committee on Government Reform, House of Representatives, 106th Congress, Second Session, April 6, 2000, "Autism: Present Challenges, Future Needs—Why the Increased Rates?" https://www.govinfo.gov/content/pkg/CHRG-106hhrg69622/html/CHRG-106hhrg69622.htm.

153 **nearly messianic terms:** Hearing Before the Committee on Government Reform, House of Representatives, 107th Congress, Second Session, June 19, 2002, "The Status of Research into Vaccine Safety and Autism," https://www.govinfo.gov/content/pkg/CHRG-107hhrg82358/html/CHRG-107hhrg82358.htm.

153 **wrote the foreword:** Jenny McCarthy, foreword to Andrew J. Wakefield, *Callous Disregard: The Truth Behind a Tragedy* (New York: Skyhorse Publishing, 2010).

NOTES

153 **traveled with a church mission:** Centers for Disease Control and Prevention (CDC), "Import-Associated Measles Outbreak—Indiana, May–June 2005," *Morbidity and Mortality Weekly Report* 54, no. 42 (2005): 1073–75, PMID: 16251862.

153 **at particular risk:** Stefan Dascalu, "Measles Epidemics in Romania: Lessons for Public Health and Future Policy," *Frontiers in Public Health* 7 (2019): 98, doi: 10.3389/fpubh.2019.00098, PMID: 31073518.

154 **attitudes about immunization:** Allison M. Kennedy and Deborah A. Gust, "Measles Outbreak Associated with a Church Congregation: A Study of Immunization Attitudes of Congregation Members," *Public Health Reports* 123, no. 2 (2008): 126–34, doi: 10.1177/003335490812300205, PMID: 18457065; Amy A. Parker et al., "Implications of a 2005 Measles Outbreak in Indiana for Sustained Elimination of Measles in the United States," *New England Journal of Medicine* 355, no. 5 (2006): 447–55, doi: 10.1056/NEJMoa060775, PMID: 16885548.

155 **more than $160,000 in costs:** Parker et al., "Implications of a 2005 Measles Outbreak in Indiana for Sustained Elimination of Measles in the United States."

155 **largest Somali American community in the United States:** Centers for Disease Control and Prevention (CDC), Somali Refugee Health Profile, https://www.cdc.gov/immigrantrefugeehealth/profiles/somali/index.html.

155 **striking change starting around 2008:** Pamala Gahr et al., "An Outbreak of Measles in an Undervaccinated Community," *Pediatrics* 134, no. 1 (2014): e220–e228, doi: 10.1542/peds.2013-4260, PMID: 24913790, fig. 1.

155 **pledged to study the problem:** Minnesota Department of Health, "Autism Spectrum Disorders Among Preschool Children Participating in the Minneapolis Public Schools Early Childhood Special Education Programs," March 2009, https://www.leg.mn.gov/docs/2009/other/090520.pdf.

156 **separate measles, mumps, and rubella vaccines:** J. B. Handley, "An Open Letter to the Somali Parents of Minnesota," *Age of Autism*, November 19, 2008, https://www.ageofautism.com/2008/11/an-open-letter.html; reprinted at Hiiraan Online, https://hiiraan.com/news2/2008/nov/an_open_letter_to_the_somali_parents_of_minnesota.aspx.

156 **"We're not an anti-vaccine movement":** Jenny McCarthy, "We're Not an Anti-Vaccine Movement . . . We're Pro-Safe Vaccine," *Frontline*, March 23, 2015, https://www.pbs.org/wgbh/frontline/article/jenny-mccarthy-were-not-an-anti-vaccine-movement-were-pro-safe-vaccine/.

156 **visited the Minnesota Somali community:** Susan Perry, "Fear and Frustration Dominated Somali Community Forum on Measles, Vaccines and Autism," MinnPost, March 28, 2011, https://www.minnpost.com/second-opinion/2011/03/fear-and-frustration-dominated-somali-community-forum-measles-vaccines-and-au/; Gahr et al., "An Outbreak of Measles in an Undervaccinated Community"; Maura Lerner, "Anti-Vaccine Doctor Meets with Somalis," *Star Tribune*, March 24, 2011, https://www.startribune.com/anti-vaccine-doctor-meets-with-somalis/118547569/; Rupa Shenoy, "Controversial Autism Researcher Tells Local Somalis Disease Is Solvable," MPR News, December 17, 2010, https://www.mprnews.org/story/2010/12/17/somali-autism.

NOTES

157 **Wakefield disavowed any role:** Lena H. Sun, "Anti-Vaccine Activists Spark a State's Worst Measles Outbreak in Decades," *Washington Post*, May 5, 2017, https://www.washingtonpost.com/national/health-science/anti-vaccine-activists-spark-a-states-worst-measles-outbreak-in-decades/2017/05/04/a1fac952-2f39-11e7-9dec-764dc781686f_story.html.

157 **reaching 42 percent among two-year-olds:** E. Banerjee et al., "Containing a Measles Outbreak in Minnesota, 2017: Methods and Challenges," *Perspectives in Public Health* 140, no. 3 (2020): 162–71, doi: 10.1177/1757913919871072, PMID: 31480896.

157 **fell below 35 percent:** Ava Kian, "Somali Children Facing the Lowest Rates of MMR Vaccinations in Minnesota History," MinnPost, August 24, 2022, https://www.minnpost.com/race-health-equity/2022/08/somali-children-facing-the-lowest-rates-of-mmr-vaccinations-in-minnesota-history/.

157 **an Amish community in Ohio:** Paul A. Gastañaduy et al., "A Measles Outbreak in an Underimmunized Amish Community in Ohio," *New England Journal of Medicine* 375, no. 14 (2016): 1343–54, doi: 10.1056/NEJMoa1602295, PMID: 27705270.

158 **At least five park employees:** R. G. Lin, R. Xia R, and N. K. Shine, "O.C. Bars Students in Measles Fight; Two Dozen Lacking Proof of Shots Must Stay Away. Five Disneyland Workers Are Among the Ill," *Los Angeles Times*, January 21, 2015, A1.

158 **barred unvaccinated children from school:** Lin, Xia, and Shine, "O.C. Bars Students in Measles Fight."

158 **confirmed 125 outbreak-linked cases:** Centers for Disease Control and Prevention (CDC), "Measles—United States, January 4–April 2, 2015," *Morbidity and Mortality Weekly Report* 64, no. 14 (2015): 373–76.

158 **Subsequent waves of infection:** R. G. Lin, "Outbreak of Measles Goes Far Beyond Disneyland; Visitors to the Park Dec. 17–20 Are Now Exposing Others, with 51 Cases Confirmed," *Los Angeles Times*, January 18, 2015, B1.

158 **focused on anti-vaccine sentiments:** Margaret K. Doll and John W. Correira, "Revisiting the 2014-15 Disneyland Measles Outbreak and Its Influence on Pediatric Vaccinations," *Human Vaccines and Immunotherapeutics* 17, no. 11 (2021): 4210–15, doi: 10.1080/21645515.2021.1972707, PMID: 34495822.

158 **"100 percent connected" to the anti-vaccine movement:** Adam Nagourney and Abby Goodnough, "Measles Cases Linked to Disneyland Rise, and Debate Over Vaccinations Intensifies," *New York Times*, January 21, 2015, A13.

158 **well known to have a vaccine hesitancy problem:** Emily Foxhall, "Parents Who Oppose Measles Vaccine Hold Firm to Their Beliefs," *Los Angeles Times*, January 25, 2015, https://www.latimes.com/local/california/la-me-measles-oc-20150126-story.html.

159 **more than nine out of ten:** Centers for Disease Control and Prevention (CDC), "Estimated Number and Percentage of Children Enrolled in Kindergarten with an Exemption to One or More Vaccines by State and the United States, Annual School Vaccination Assessment Report, 2014–15 School Year," https://www.cdc.gov/vaccines/imz-managers/coverage/schoolvaxview/data-reports/index.html.

159 **nonmedical exemptions had been increasing:** Jennifer L. Richards et al., "Nonmedical Exemptions to Immunization Requirements in California: a 16-Year Longitudinal

NOTES

Analysis of Trends and Associated Community Factors," *Vaccine* 31, no. 29 (2013): 3009–13, PMID: 23664998; Saad B. Omer et al., "Vaccination Policies and Rates of Exemption from Immunization, 2005–2011," *New England Journal of Medicine* 367, no. 12 (2012): 1170–71, doi: 10.1056/NEJMc1209037, PMID: 22992099.

159 **nonmedical exemptions were geographically clustered:** Margaret Carrel and Patrick Bitterman, "Personal Belief Exemptions to Vaccination in California: A Spatial Analysis," *Pediatrics* 136, no. 1 (2015): 80–88, doi: 10.1542/peds.2015-0831, PMID: 26034242.

159 **more than a third of children:** Foxhall, "Parents Who Oppose Measles Vaccine Hold Firm to Their Beliefs."

159 **wealthy, predominantly white areas:** Carrel and Bitterman, "Personal Belief Exemptions to Vaccination in California: A Spatial Analysis."

159 **attracted vaccine-hesitant families:** Julia M. Brennan et al., "Trends in Personal Belief Exemption Rates Among Alternative Private Schools: Waldorf, Montessori, and Holistic Kindergartens in California, 2000–2014," *American Journal of Public Health* 107, no. 1 (2017): 108–112, doi: 10.2105/AJPH.2016.303498, PMID: 27854520.

159 **more than 40 percent of children:** Foxhall, "Parents Who Oppose Measles Vaccine Hold Firm to Their Beliefs."

159 **Los Altos, one of the wealthiest towns:** KJ Dell'Antonia, "The Waldorf Way: No Tests, No Grades, No Shots?" *Slate*, November 1, 2011, https://slate.com/human-interest/2011/11/who-dares-confront-the-parents-of-the-bay-area-waldorf-school-with-the-23-vaccination-rate.html; Megan McArdle, "A Shocking Chart on Vaccination," *The Atlantic*, October 31, 2011, https://www.theatlantic.com/health/archive/2011/10/a-shocking-chart-on-vaccination/247651/.

159 **appealing to vaccine-hesitant parents:** Kimiko de Freytas-Tamura, "Bastion of Anti-Vaccine Fervor: Progressive Waldorf Schools," *New York Times*, June 13, 2019, https://www.nytimes.com/2019/06/13/nyregion/measles-outbreak-new-york.html; Elisa J. Sobo, "Social Cultivation of Vaccine Refusal and Delay Among Waldorf (Steiner) School Parents," *Medical Anthropology Quarterly* 29, no. 3 (2015): 381–99, doi: 10.1111/maq.12214, PMID: 25847214.

159 **every kindergarten in the state:** Matthew Bloch, Josh Keller, and Haeyoun Park, "Vaccination Rates for Every Kindergarten in California," *New York Times*, February 6, 2015, https://www.nytimes.com/interactive/2015/02/06/us/california-measles-vaccines-map.html.

160 **"not a lot of cases":** Nagourney and Goodnough, "Measles Cases Linked to Disneyland Rise, and Debate Over Vaccinations Intensifies."

160 **Measles-related social media activity spiked:** A. Demetri Pananos et al., "Critical Dynamics in Population Vaccinating Behavior," *Proceedings of the National Academy of Sciences of the USA* 114, no. 52 (2017): 13762–67, doi: 10.1073/pnas.1704093114, PMID: 29229821.

160 **made up of pro-vaccine posts:** Michael S. Deiner et al., "Facebook and Twitter Vaccine Sentiment in Response to Measles Outbreaks," *Health Informatics Journal* 25, no. 3 (2019): 1116–32, doi: 10.1177/1460458217740723, PMID: 29148313.

NOTES

160 **"ignorant and self-absorbed rejection of science"**: "Editorial: The Disneyland Outbreak," *Los Angeles Times*, January 16, 2015, A20.

160 **bring about a "Disneyland effect"**: Joshua M. Sharfstein, "Of Mouse and Measles," *JAMA* 313, no. 15 (2015): 1504–5, doi: 10.1001/jama.2015.1135, PMID: 25898033; Doll and Correira, "Revisiting the 2014-15 Disneyland Measles Outbreak and Its Influence on Pediatric Vaccinations"; Margaret K. Doll, Samuel D. Weitzen, and Kathryn T. Morrison, "Trends in the Uptake of Pediatric Measles-Containing Vaccine in the United States: A Disneyland Effect?" *Vaccine* 39, no. 2 (2021): 357–63, doi: 10.1016/j.vaccine.2020.11.048, PMID: 33288341.

160 **"My parents are idiots"**: Emily Flake, "Daily Cartoon: Monday, February 2nd," *New Yorker*, February 2, 2015, https://www.newyorker.com/cartoons/daily-cartoon/daily-cartoon-monday-february-2nd-measles-disneyland.

160 **totaled 147 confirmed cases:** Nakia S. Clemmons et al., "Measles—United States, January 4–April 2, 2015," *Morbidity and Mortality Weekly Report* 64, no. 14 (2015): 373–76, PMID: 25879894; "Measles Outbreak Traced to Disneyland Is Declared Over," NBC News, April 17, 2015, https://www.nbcnews.com/storyline/measles-outbreak/measles-outbreak-traced-disneyland-declared-over-n343686.

160 **seeded a separate two-month-long outbreak:** L. Sherrard et al., "Measles Surveillance in Canada: 2015," *Canadian Communicable Disease Report* 42, no. 7 (2016): 139–45, doi: 10.14745/ccdr.v42i07a01, PMID: 29770019.

161 **elimination of nonmedical exemptions:** Senate Bill 277—California Legislative Information, https://leginfo.legislature.ca.gov/faces/billNavClient.xhtml?bill_id=201520160SB277.

161 **meet with a health care provider:** Lisa Aliferis, "New Requirement for Vaccine Exemption Passed by Senate," KQED, August 23, 2012, https://www.kqed.org/stateofhealth/8228/new-requirement-for-vaccine-exemption-passed-by-senate.

161 **threatened to pull their children out of school:** Tracy Seipel, "Foes Vow to Home-School: Opposition Heats Up as Bill Heads to Vote by Education Committee," *San Jose Mercury News*, April 14, 2015, B1.

161 **lobbied legislators:** Melanie Mason, "Chiropractors Oppose Vaccines Rule; Group Lobbies to Keep Children's Exemption Based on Parents' Personal Beliefs," *Los Angeles Times*, March 5, 2015, B4.

161 **apologized for the comparison:** Patrick McGreevy, "RFK Jr. Regrets Use of Word in Speech; He Apologizes for Using 'Holocaust' to Describe Health Effects of Vaccine Programs," *Los Angeles Times*, April 14, 2015, B5.

161 **once again he apologized:** Andrew Jeong, "Robert F. Kennedy Jr. Apologizes for Saying the Unvaccinated Have Less Freedom Than Anne Frank Did," *Washington Post*, January 25, 2022, https://www.washingtonpost.com/nation/2022/01/25/rfk-kennedy-jr-anne-frank-anti-vax/.

161 **Senator Pan received death threats:** "Death Threats Prompt Increased Security for California Vaccination Bill Author," CBS Sacramento, April 14, 2015, https://sacramento.cbslocal.com/2015/04/14/death-threats-prompt-increased-security-for-california-vaccination-bill-author/.

NOTES

161 **signed into law:** Senate Bill 277—California Legislative Information, https://leginfo.leg islature.ca.gov/faces/billNavClient.xhtml?bill_id=201520160SB277; Maggie Fox, "California Governor Signs Tough New Vaccine Law," NBC News, June 30, 2015, https://www.nbcnews.com/health/health-news/california-governor-signs-tough-new-vaccine-law-n384556.

161 **vaccination rates in California schools increased significantly:** Paul L. Delamater et al., "Elimination of Nonmedical Immunization Exemptions in California and School-Entry Vaccine Status," *Pediatrics* 143, no. 6 (2019): e20183301, doi: 10.1542/peds.2018-3301, PMID: 31113831.

161 **tripled in the year following SB 277:** Delamater et al., "Elimination of Nonmedical Immunization Exemptions in California and School-Entry Vaccine Status"; Paul L. Delamater, Timothy F. Leslie, and Y. Tony Yang, "Change in Medical Exemptions from Immunization in California After Elimination of Personal Belief Exemptions," *JAMA* 318, no. 9 (2017): 863–64, doi: 10.1001/jama.2017.9242, PMID: 28873152.

161 **legitimate medical exemptions had been rare:** Delamater, Leslie, and Yang, "Change in Medical Exemptions from Immunization in California After Elimination of Personal Belief Exemptions."

161 **similar clustering emerged:** Ashley Gromis and Ka-Yuet Liu, "The Emergence of Spatial Clustering in Medical Vaccine Exemptions Following California Senate Bill 277, 2015–2018," *American Journal of Public Health* 110, no. 7 (2020): 1084–91, doi: 10.2105/AJPH.2020.305607, PMID: 32437268.

161 **medical exemptions without a legitimate indication:** Ashley Gromis and Ka-Yuet Liu, "Spatial Clustering of Vaccine Exemptions on the Risk of a Measles Outbreak," *Pediatrics* 149, no. 1 (2022): e2021050971, doi: 10.1542/peds.2021-050971, PMID: 34866158.

162 **backed up this hypothesis:** Salini Mohanty et al., "California's Senate Bill 277: Local Health Jurisdictions' Experiences with the Elimination of Nonmedical Vaccine Exemptions," *American Journal of Public Health* 109, no. 1 (2019): 96–101, doi: 10.2105/AJPH.2018.304768, PMID: 30495995; Salini Mohanty et al., "Experiences with Medical Exemptions After a Change in Vaccine Exemption Policy in California," *Pediatrics* 142, no. 5 (2018): e20181051, doi: 10.1542/peds.2018-1051, PMID: 30373910.

162 **despite intensive lobbying:** Adeel Hassan, "Here Is What Jessica Biel Opposes in California's Vaccine Bill," *New York Times*, June 13, 2019, https://www.nytimes.com/2019/06/13/us/california-vaccine-bill.html; Jordan Julian, "Comedian Rob Schneider Has Become Hollywood's Loudest (and Wackiest) Anti-Vaxxer," *The Daily Beast*, July 27, 2019, https://www.thedailybeast.com/comedian-rob-schneider-has-become-hollywoods-loudest-and-wackiest-anti-vaxxer.

162 **Biel posted on Instagram:** Bruce Haring, "Jessica Biel Adds Her Voice Against California Bill Limiting Medical Exemptions for Vaccinations; Later Clarifies Position," *Deadline*, June 12, 2019, https://deadline.com/2019/06/jessica-biel-california-sb-276-vaccinations-bill-1202631983/.

162 **menstrual cup containing blood:** Angela Hart and Colby Bermel, "Protester Throws Apparent Blood at Legislators, Shutting Down California Senate," Politico, September 13,

NOTES

2019, https://www.politico.com/states/california/story/2019/09/13/protester-throws-red-liquid-at-legislators-shutting-down-california-senate-1188537.

162 **physically assaulted Pan:** Chris Jennewein, "State Senator Who Authored Vaccination Bills Assaulted by Anti-Vaxxer at Capitol," *Times of San Diego*, August 21, 2019, https://timesofsandiego.com/politics/2019/08/21/state-senator-who-authored-vaccination-bills-assaulted-by-anti-vaxxer-at-capitol.

162 **increased substantially in the majority of states:** Jacqueline K. Olive et al., "The State of the Anti-Vaccine Movement in the United States: A Focused Examination of Nonmedical Exemptions in States and Counties," *PLoS Medicine* 15, no. 6 (2018): e1002578, doi: 10.1371/journal.pmed.1002578, PMID: 29894470.

162 **puts those areas at increased risk:** Olive et al., "The State of the Anti-Vaccine Movement in the United States: A Focused Examination of Nonmedical Exemptions in States and Counties."

163 **medical exemptions had already climbed significantly:** Julian Shen-Berro, "New York's Private Schools Are Gaming Vaccine Exemptions in 'Obvious' Fraud," Politico, October 1, 2022, https://www.politico.com/news/2022/10/01/state-nixed-religious-vaccine-exemptions-medical-exemptions-up-00059089.

163 **"I really think it does something to the children":** Josh Hafenbrack, "Trump: Autism Linked to Child Vaccinations," *South Florida Sun Sentinel*, December 28, 2007, https://www.sun-sentinel.com/sfl-mtblog-2007-12-trump_autism_linked_to_child_v-story.html.

163 **spacing out the vaccines over a long period of time:** Hafenbrack, "Trump: Autism Linked to Child Vaccinations."

163 **"Massive combined inoculations":** Donald J. Trump, https://twitter.com/realdonaldtrump/status/238717783007977473.

163 **"Autism WAY UP":** Donald J. Trump, https://twitter.com/realdonaldtrump/status/449329067192762368.

163 **"extremely educated" on the issues:** Zack Kopplin, "Trump Met with Prominent Anti-Vaccine Activists During Campaign," *Science*, November 18, 2016, doi: 10.1126/science.aal0407, https://www.science.org/content/article/trump-met-prominent-anti-vaccine-activists-during-campaign.

163 **Wakefield said that he had been pleased:** Rebecca Robbins, "Meeting with Trump Emboldens Anti-Vaccine Activists, Who See an Ally in the Oval Office," STAT, November 30, 2016, https://www.statnews.com/2016/11/30/donald-trump-vaccines-policy/.

163 **calling for a "huge shake-up" at the CDC:** Sarah Boseley, "How Disgraced Anti-Vaxxer Andrew Wakefield Was Embraced by Trump's America," *The Guardian*, July 18, 2018, https://www.theguardian.com/society/2018/jul/18/how-disgraced-anti-vaxxer-andrew-wakefield-was-embraced-by-trumps-america; Casey Ross, "Andrew Wakefield Appearance at Trump Inaugural Ball Triggers Social Media Backlash," STAT, January 21, 2017, https://www.statnews.com/2017/01/21/andrew-wakefield-trump-inaugural-ball/.

164 **clear message to the anti-vaccine movement:** Brian Deer and Josh Glancy, "Trump Gives Anti-Vaccine Zealots a Shot in the Arm," *The Times (London)* January 29, 2017,

NOTES

https://www.thetimes.co.uk/article/trump-gives-anti-vaccine-zealots-a-shot-in-the-arm-lwtbhdcgq; Boseley, "How Disgraced Anti-Vaxxer Andrew Wakefield Was Embraced by Trump's America"; Ruth Graham, "Vaccine Skeptics Are Excited About Donald Trump's Presidency," Slate, November 30, 2016, https://slate.com/human-interest/2016/11/vaccine-skeptic-andrew-wakefield-is-excited-about-donald-trumps-presidency.html.

CHAPTER 10: BANNER YEARS

165 **loss of progress in places where transmission:** Alya Dabbagh et al., "Progress Toward Regional Measles Elimination—Worldwide, 2000–2017," *Morbidity and Mortality Weekly Report* 67, no. 47 (2018): 1323–29, doi: 10.15585/mmwr.mm6747a6, PMID: 30496160.

165 **measles transmission returned:** Dabbagh et al., "Progress Toward Regional Measles Elimination—Worldwide, 2000–2017."

165 **vaccination rates of Ukrainian children:** Meredith Wadman, "Measles Cases Have Tripled in Europe, Fueled by Ukrainian Outbreak," *Science*, February 12, 2019, doi: 10.1126/science.aaw9303, https://www.science.org/content/article/measles-cases-have-tripled-europe-fueled-ukrainian-outbreak.

166 **unholy explosion of cases:** Center for Disease Control and Prevention (CDC), "Measles Cases and Outbreaks," https://www.cdc.gov/measles/cases-outbreaks.html.

167 **hotbeds of contagion:** Albert E. Barskey et al., "Mumps Outbreak in Orthodox Jewish Communities in the United States," *New England Journal of Medicine* 367, no. 18 (2012): 1704–13, doi: 10.1056/NEJMoa1202865, PMID: 23113481.

167 **most of the affected individuals had received two doses:** Barskey et al., "Mumps Outbreak in Orthodox Jewish Communities in the United States."

167 **highly vaccinated but close-knit groups:** Gustavo H. Dayan et al., "Recent Resurgence of Mumps in the United States," *New England Journal of Medicine* 358, no. 15 (2008): 1580–89, doi: 10.1056/NEJMoa0706589, PMID: 18403766; Joseph A. Lewnard and Yonatan H. Grad, "Vaccine Waning and Mumps Re-emergence in the United States," *Science Translational Medicine* 10, no. 433 (2018): eaao5945, doi: 10.1126/scitranslmed.aao5945, PMID: 29563321; Shirlee Wohl et al., "Combining Genomics and Epidemiology to Track Mumps Virus Transmission in the United States," *PLoS Biology* 18, no. 2 (2020): e3000611, doi: 10.1371/journal.pbio.3000611, PMID: 32045407.

168 **clear hot spots:** Barskey et al., "Mumps Outbreak in Orthodox Jewish Communities in the United States," fig. 3.

168 **returned to Borough Park:** Centers for Disease Control and Prevention (CDC), "Notes from the Field: Measles Outbreak Among Members of a Religious Community—Brooklyn, New York, March–June 2013," *Morbidity and Mortality Weekly Report* 62, no. 36 (2013): 752–3, PMID: 24025758; Jennifer B. Rosen et al., "Public Health Consequences of a 2013 Measles Outbreak in New York City," *JAMA Pediatrics* 172, no. 9 (2018): 811–17, doi: 10.1001/jamapediatrics.2018.1024, PMID: 30073293.

169 **unvaccinated status of all the cases:** CDC, "Notes from the Field: Measles Outbreak Among Members of a Religious Community—Brooklyn, New York, March–June 2013."

NOTES

169 **developed severe herpes infections:** Centers for Disease Control and Prevention (CDC), "Neonatal Herpes Simplex Virus Infection Following Jewish Ritual Circumcisions That Included Direct Orogenital Suction—New York City, 2000–2011," *Morbidity and Mortality Weekly Report* 61, no. 22 (2012): 405–409, PMID: 22672975.

169 **outrage and threats of lawsuits:** Sharon Otterman, "Denouncing City's Move to Regulate Circumcision," *New York Times*, September 12, 2012, https://www.nytimes.com/2012/09/13/nyregion/regulation-of-circumcision-method-divides-some-jews-in-new-york.html.

170 **continued to perform the procedure:** Paul A. Offit, "Ban the Ritual That Can Kill Jewish Newborns," *The Daily Beast*, May 5, 2017, https://www.thedailybeast.com/ban-the-ritual-that-can-kill-jewish-newborns.

170 **established a hotline:** Amanda Schaffer, "Fear, Misinformation, and Measles Spread in Brooklyn," *Wired*, June 24, 2019, https://www.wired.com/story/fear-misinformation-measles-spread-in-brooklyn/; Gwynne Hogan, "Misinformation Hotline Stokes Fear of Vaccines in Ultra-Orthodox Community," Gothamist, March 12, 2019, https://gothamist.com/news/misinformation-hotline-stokes-fear-of-vaccines-in-ultra-orthodox-community.

170 **identified in other reporting:** Brandy Zadrozny, "Brooklyn Measles Outbreak: How a Glossy Booklet Spread Anti-Vaccine Messages in Orthodox Jewish Communities," NBC News, April 12, 2019, https://www.nbcnews.com/news/us-news/brooklyn-measles-outbreak-how-glossy-booklet-spread-anti-vaccine-messages-n993596; Hogan, "Misinformation Hotline Stokes Fear of Vaccines in Ultra-Orthodox Community."

170 **descent into anti-vaccine conspiracy theories:** Schaffer, "Fear, Misinformation, and Measles Spread in Brooklyn."

171 **cites as an early influence:** Schaffer, "Fear, Misinformation, and Measles Spread in Brooklyn."

172 **mass mailings to Jewish communities in other cities:** Tory Tabachnik, "Anonymous Anti-Vaxxers Push Propaganda on Local Orthodox Community," *Pittsburgh Jewish Chronicle*, January 31, 2018, https://jewishchronicle.timesofisrael.com/anonymous-anti-vaxxers-push-propaganda-on-local-orthodox-community/.

172 **ten times the rate:** Jennifer Rosen, "Measles Outbreak New York City, 2018–19," https://cheac.org/wp-content/uploads/2019/10/Measles_CHEAC_Oct2019.pdf; Manisha Patel et al., "Measles—Maintaining Disease Elimination and Enhancing Vaccine Confidence," *CDC Public Health Grand Rounds*, February 18, 2020, https://www.cdc.gov/grand-rounds/pp/2020/20200218-measles-elimination-vaccine-H.pdf, Video: https://www.youtube.com/watch?v=YJPabiGf1TE.

172 **under 80 percent:** Rosen et al., "Public Health Consequences of a 2013 Measles Outbreak in New York City."

172 **nearly one in four schoolchildren:** Robert McDonald, et al., "Notes from the Field: Measles Outbreaks from Imported Cases in Orthodox Jewish Communities—New York and New Jersey, 2018–2019," *Morbidity and Mortality Weekly Report* 68, no. 19 (2019): 444–45, doi: 10.15585/mmwr.mm6819a4, PMID: 31095533.

NOTES

173 **took off in larger Israeli cities:** Marcy Oster, "Expert: 'Israel's Measles Outbreak Began at Uman Hasidic Pilgrimage,'" *Times of Israel*, April 4, 2019, https://www.timesofisrael.com/expert-israels-measles-outbreak-began-at-uman-hasidic-pilgrimage/; Donald G. McNeil, "Scientists Thought They Had Measles Cornered. They Were Wrong," *New York Times*, April 3, 2019, https://www.nytimes.com/2019/04/03/health/measles-outbreaks-ukraine-israel.html; Matanelle Salama et al., "A Measles Outbreak in the Tel Aviv District, Israel, 2018–2019," *Clinical Infectious Diseases* 72, no. 9 (2021): 1649–56, doi: 10.1093/cid/ciaa931, PMID: 32619227; Chen Stein-Zamir, Nitza Abramson, and Hanna Shoob, "Notes from the Field: Large Measles Outbreak in Orthodox Jewish Communities—Jerusalem District, Israel, 2018–2019," *Morbidity and Mortality Weekly Report* 69, no. 18 (2020): 562–63, doi: 10.15585/mmwr.mm6918a3, PMID: 32379730.

173 **rates in Orthodox enclaves were much lower:** Salama et al., "A Measles Outbreak in the Tel Aviv District, Israel, 2018–2019"; Stein-Zamir, Abramson, and Shoob, "Notes from the Field: Large Measles Outbreak in Orthodox Jewish Communities—Jerusalem District, Israel, 2018–2019."

173 **attended a crowded synagogue:** Nick Paumgarten, "The Message of Measles," *The New Yorker*, September 2, 2019, https://www.newyorker.com/magazine/2019/09/02/the-message-of-measles.

173 **printed up flyers, held vaccination drives, went door to door:** Sharon Otterman, "New York Confronts Its Worst Measles Outbreak in Decades," *New York Times*, January 17, 2019, https://www.nytimes.com/2019/01/17/nyregion/measles-outbreak-jews-nyc.html.

173 **allow their unvaccinated children to return:** Reis Thebault, "The Parents of More than Three Dozen Unvaccinated Kids Want Them Back in School. A Judge Said No," *Washington Post*, March 14, 2019, https://www.washingtonpost.com/health/2019/03/14/parents-unvaccinated-kids-wanted-them-back-school-now-judge-said-no/; Lisa Miller, "Measles for the One Percent," *The Cut*, May 29, 2019, https://www.thecut.com/2019/05/measles-for-the-one-percent.html.

174 **challenges to vaccine mandates:** Wendy K. Mariner, George J. Annas, and Leonard H. Glantz, "Jacobson v. Massachusetts: It's Not Your Great-Great-Grandfather's Public Health Law," *American Journal of Public Health* 95, no. 4 (2005): 581–90, doi: 10.2105/AJPH.2004.055160, PMID: 15798113; Wendy E. Parmet, et al. "Individual Rights Versus the Public's Health—100 Years After Jacobson v. Massachusetts," *New England Journal of Medicine* 352, no. 7 (2005): 652–54, doi: 10.1056/NEJMp048209, PMID: 15716558.

174 **infected more than twenty others:** Tyler Pager, "Measles Outbreak: 1 Student Got 21 Others Sick," *New York Times*, March 7, 2019, https://www.nytimes.com/2019/03/07/nyregion/measles-outbreak-vaccine-nyc.html.

174 **index case for that outbreak:** Alyssa Carlson et al., "Notes from the Field: Community Outbreak of Measles—Clark County, Washington, 2018–2019," *Morbidity and Mortality Weekly Report* 68, no. 19 (2019): 446–47, doi: 10.15585/mmwr.mm6819a5, PMID: 31095534.

174 **reported fifty-five measles deaths among children:** Jason Gutierrez, "Measles Outbreak in Philippines Spreads Beyond Capital," *New York Times*, February 7, 2019, https://www.nytimes.com/2019/02/07/world/asia/philippines-measles-outbreak.html.

NOTES

174 **top ten threats to global health:** Anne Gulland, "Air Pollution, Obesity and Vaccine Hesitancy Climb WHO List of Global Health Threats," *The Telegraph*, January 15, 2019, https://www.telegraph.co.uk/global-health/climate-and-people/air-pollution-obesity-vaccine-hesitancy-climb-list-global-health/; World Health Organization, "Ten Threats to Global Health in 2019," https://www.who.int/news-room/spotlight/ten-threats-to-global-health-in-2019.

174 **invited to testify:** James Doubek, "18-Year-Old Testifies About Getting Vaccinated Despite Mother's Anti-Vaccine Beliefs," NPR News, March 6, 2019, https://www.npr.org/2019/03/06/700617424/18-year-old-testifies-about-getting-vaccinated-despite-mothers-anti-vaccine-beli; Hearing of the Committee on Health, Education, Labor, and Pensions, United States Senate, 116th Congress, First Session on Examining Vaccines, Focusing on Preventable Disease Outbreaks, March 5, 2019, Vaccines Save Lives: What Is Driving Preventable Disease Outbreaks? https://www.govinfo.gov/content/pkg/CHRG-116shrg41391/pdf/CHRG-116shrg41391.pdf. Video: https://www.help.senate.gov/hearings/vaccines-save-lives-what-is-driving-preventable-disease-outbreaks.

174 **plans to begin to combat misinformation:** Christina Caron, "Facebook Announces Plan to Curb Vaccine Misinformation," *New York Times*, March 8, 2019, B9, https://www.nytimes.com/2019/03/07/technology/facebook-anti-vaccine-misinformation.html.

174 **denouncing the purported vaccines and autism link:** Brett P. Giroir, Robert R. Redfield, and Jerome M. Adams, "This Is the Truth About Vaccines," *New York Times*, March 6, 2019, https://www.nytimes.com/2019/03/06/opinion/vaccines-autism-flu.html.

175 **declared a state of emergency:** E. J. Day, "Declaration of a Local State of Emergency for Rockland County—Amendment No. 2," March 26, 2019, https://rocklandgov.com/files/3815/5432/3776/doe_4-2.pdf, press release, http://rocklandgov.com/departments/county-executive/press-releases/2019-press-releases/state-of-emergency-declared/; Matthew S. Schwartz, "N.Y. Suburb Declares Measles Emergency, Bars Unvaccinated Minors from Public Places," NPR News, March 27, 2019, https://www.npr.org/2019/03/27/707095754/ny-suburb-declares-measles-emergency-bars-unvaccinated-minors-from-public-places; Michael Gold and Tyler Pager, "New York Suburb Declares Measles Emergency, Barring Unvaccinated Children from Public," *New York Times*, March 26, 2019, https://www.nytimes.com/2019/03/26/nyregion/measles-outbreak-rockland-county.html.

175 **intended as an "attention grabber":** Schwartz, "N.Y. Suburb Declares Measles Emergency, Bars Unvaccinated Minors from Public Places."

175 **distrust in the community might deepen:** Gold and Pager, "New York Suburb Declares Measles Emergency, Barring Unvaccinated Children from Public."

175 **Mayor de Blasio declared a public health emergency:** Gold and Pager, "New York Suburb Declares Measles Emergency, Barring Unvaccinated Children from Public"; Oxiris Barbot, "Order of the Commissioner," https://www1.nyc.gov/assets/doh/downloads/pdf/press/2019/emergency-orders-measles.pdf; Julie D. Cantor, "Mandatory Measles Vaccination in New York City—Reflections on a Bold Experiment," *New England Journal of Medicine* 381, no. 2 (2019): 101–103, doi: 10.1056/NEJMp1905941, PMID: 31167046.

NOTES

175 **PEACH held a four-hour conference call:** Tyler Pager, "'Monkey, Rat and Pig DNA': How Misinformation Is Driving the Measles Outbreak Among Ultra-Orthodox Jews," *New York Times*, April 9, 2019, https://www.nytimes.com/2019/04/09/nyregion/jews-measles-vaccination.html.

175 **Orthodox families received robocalls:** Pager, "'Monkey, Rat and Pig DNA': How Misinformation Is Driving the Measles Outbreak Among Ultra-Orthodox Jews."

175 **shut down a preschool program:** Tyler Pager, "Measles Outbreak: Yeshiva's Preschool Program Is Closed by New York City Health Officials," *New York Times*, April 15, 2019, https://www.nytimes.com/2019/04/15/nyregion/measles-nyc-yeshiva-closing.html.

176 **issues in divorce proceedings:** Lela Moore, "A Pregnant Woman Avoids Transit, Parents Battle in Court and Other Tales of Measles Anxiety," *New York Times*, May 3, 2019, https://www.nytimes.com/2019/05/03/reader-center/measles-outbreak-america.html.

176 **Incidents of anti-Semitism connected with measles rose:** Emma Green, "Measles Can Be Contained. Anti-Semitism Cannot," *The Atlantic*, May 25, 2019, https://www.theatlantic.com/politics/archive/2019/05/orthodox-jews-face-anti-semitism-after-measles-outbreak/590311/.

176 **deliberate misinformation:** Moshe Friedman, "My Fellow Hasidic Jews Are Making a Terrible Mistake About Vaccinations," *New York Times*, April 23, 2019, https://www.nytimes.com/2019/04/23/opinion/my-fellow-hasidic-jews-are-making-a-terrible-mistake-about-vaccinations.html.

176 **fighting a disease that we already knew how to prevent:** Perri Klass, "Measles Now," *New York Times*, April 15, 2019, https://www.nytimes.com/2019/04/15/well/family/measles-now.html.

176 **requiring nearly a thousand college students:** Soumya Karlamangla, "Will Measles Orders Quell the Outbreak? More Than 1,000 from L.A. Universities Told to Remain at Home," *Los Angeles Times*, April 27, 2019, A1.

176 **attempting to land with an active measles case on board:** Kate Feldman, "Scientology Cruise Ship Is Quarantined," *Los Angeles Times*, May 3, 2019, A-4.

176 **vaccinated without their mother's knowledge:** Kwame Anthony Appiah, "Can I Get My Anti-Vaxx Sister's Kids Vaccinated?" *New York Times*, May 14, 2019, https://www.nytimes.com/2019/05/14/magazine/can-i-get-my-anti-vaxx-sisters-kids-vaccinated.html.

176 **local anti-vaccine personalities:** Kimiko de Freytas-Tamura, "Despite Measles Warnings, Anti-Vaccine Rally Draws Hundreds of Ultra-Orthodox Jews," *New York Times*, May 14, 2019, https://www.nytimes.com/2019/05/14/nyregion/measles-vaccine-orthodox-jews.html; Beth Mole, "Andrew Wakefield, Others Hold Anti-Vaccine Rally Amid Raging Measles Outbreaks," *Ars Technica*, May 14, 2019, https://arstechnica.com/science/2019/05/andrew-wakefield-others-hold-anti-vaccine-rally-amid-raging-measles-outbreaks/.

176 **falsely claimed that bad lots of vaccine:** de Freytas-Tamura, "Despite Measles Warnings, Anti-Vaccine Rally Draws Hundreds of Ultra-Orthodox Jews."

177 **"the campaign against us has been successful":** de Freytas-Tamura, "Despite Measles Warnings, Anti-Vaccine Rally Draws Hundreds of Ultra-Orthodox Jews."

NOTES

177 **claimed that Mayor de Blasio was German:** Mole, "Andrew Wakefield, Others Hold Anti-Vaccine Rally Amid Raging Measles Outbreaks."

177 **addressed the crowd via videoconference:** Mole, "Andrew Wakefield, Others Hold Anti-Vaccine Rally Amid Raging Measles Outbreaks"; de Freytas-Tamura, "Despite Measles Warnings, Anti-Vaccine Rally Draws Hundreds of Ultra-Orthodox Jews."

177 **join a group text chat of about forty mothers:** Amanda Schaffer, "Amid a Measles Outbreak, an Ultra-Orthodox Nurse Fights Vaccination Fears in Her Community," *The New Yorker*, January 25, 2019, https://www.newyorker.com/news/as-told-to/amid-a-measles-outbreak-an-ultra-orthodox-nurse-fights-vaccination-fears-in-her-community; Lior Zaltzman, "These Amazing Orthodox Nurses Are Fighting Measles in Their Community," Kveller, April 3, 2019. https://www.kveller.com/these-amazing-orthodox-nurses-are-fighting-measles-in-their-community/; Blima Marcus, author interview, August 14, 2023.

177 **form a task force on vaccines:** Blima Marcus, "A Nursing Approach to the Largest Measles Outbreak in Recent U.S. History: Lessons Learned Battling Homegrown Vaccine Hesitancy," *OJIN: The Online Journal of Issues in Nursing* 25, no. 1 (2020), https://doi.org/10.3912/OJIN.Vol25No01Man03; EMES Initiative, https://emesinitiative.org.

177 **discussions were civil and fact-based:** Schaffer, "Fear, Misinformation, and Measles Spread in Brooklyn."

178 **particularly vexed by the PEACH booklet:** Gwynne Hogan, "How Orthodox Jewish Nurses Are Fighting 'Anti-Vaccination Propaganda' Targeting Their Community," Gothamist, March 26, 2019, https://gothamist.com/news/how-orthodox-jewish-nurses-are-fighting-anti-vaccination-propaganda-targeting-their-community; Schaffer, "Fear, Misinformation, and Measles Spread in Brooklyn."

178 **abridged version of *PIE* focused on the MMR vaccine:** Marcus, "A Nursing Approach to the Largest Measles Outbreak in Recent U.S. History: Lessons Learned Battling Homegrown Vaccine Hesitancy"; EMES Initiative, A Slice of PIE, https://www.nyc.gov/assets/doh/downloads/pdf/imm/a-slice-of-pie.pdf.

178 **more than 185,000 doses:** Jane R. Zucker et al., "Consequences of Undervaccination—Measles Outbreak, New York City, 2018–2019," *New England Journal of Medicine* 382, no. 11 (2020): 1009–17, doi: 10.1056/NEJMoa1912514, PMID: 32160662.

179 **published papers in nursing journals:** Lena H. Sun, "Nurses Are Teaching Doctors How to Treat Anti-Vaccine Fears and Myths," *Washington Post*, July 21, 2019, https://www.washingtonpost.com/health/nurses-are-teaching-doctors-how-to-treat-anti-vaccine-fears-and-myths/2019/07/15/79632f9e-a3fd-11e9-bd56-eac6bb02d01d_story.html; Marcus, "A Nursing Approach to the Largest Measles Outbreak in Recent U.S. History: Lessons Learned Battling Homegrown Vaccine Hesitancy"; Lindsey Danielson, Blima Marcus, and Lori Boyle, "Special Feature: Countering Vaccine Misinformation," *American Journal of Nursing* 119, no. 10 (2019): 50–55, doi: 10.1097/01.NAJ.0000586176.77841.86, PMID: 31567253.

179 **EMES held another fair:** Bobby Allyn, "New York Ends Religious Exemptions for Required Vaccines," NPR News, June 13, 2019, https://www.npr.org/2019/06/13/732501865/new

NOTES

-york-advances-bill-ending-religious-exemptions-for-vaccines-amid-health-cris; New York State Assembly, Bill A02371 Summary, https://nyassembly.gov/leg/?default_fld=&leg_video=&bn=A02371&term=2019&Summary=Y&Actions=Y&Committee%26nbspVotes=Y&Floor%26nbspVotes=Y&Memo=Y&Text=Y.

179 **letter went out from five hundred Orthodox physicians:** Marcy Oster, "500 US Doctors Sign Letter Urging Orthodox Jews to Vaccinate," *Times of Israel*, April 17, 2019, https://www.timesofisrael.com/500-doctors-sign-letter-urging-orthodox-jews-to-vaccinate/.

179 **Numerous rabbis provided pro-vaccine messages:** Y. Hoffman, "Rav Moshe Sternbuch Writes Letter to Rav Malkiel Kotler About the Halachic Requirement to Vaccinate," *Yeshiva World News*, November 27, 2018, https://www.theyeshivaworld.com/news/general/1631188/rav-moshe-sternbuch-writes-letter-to-rav-malkiel-kotler-about-the-halachic-requirement-to-vaccinate.html.

179 **outside the framework that they would generally use:** Ben Kasstan, "If a Rabbi Did Say 'You Have to Vaccinate,' We Wouldn't: Unveiling the Secular Logics of Religious Exemption and Opposition to Vaccination," *Social Science & Medicine* 280 (2021): 114052, doi: 10.1016/j.socscimed.2021.114052, PMID: 34051560; Yael Keshet and Ariela Popper-Giveon, "'I Took the Trouble to Make Inquiries, So I Refuse to Accept Your Instructions': Religious Authority and Vaccine Hesitancy Among Ultra-Orthodox Jewish Mothers in Israel," *Journal of Religion and Health* 60, no. 3 (2021): 1992–2006, doi: 10.1007/s10943-020-01122-4, PMID: 33389435.

180 **Rockland had 312 confirmed measles cases:** E. Day, press release: "Measles Outbreak Declared Over in Rockland," http://rocklandgov.com/departments/county-executive/press-releases/2019-press-releases/measles-outbreak-declared-over-in-rockland/.

180 **had cost the city about $400,000:** Rosen et al., "Public Health Consequences of a 2013 Measles Outbreak in New York City."

180 **more than $8 million:** Zucker et al., "Consequences of Undervaccination—Measles Outbreak, New York City, 2018–2019."

180 **received no doses of measles-containing vaccines:** Zucker et al., "Consequences of Undervaccination—Measles Outbreak, New York City, 2018–2019."

181 **a highly publicized 2018 case:** Catherine Graue and Michael Walsh, "Samoa Recalls Vaccines, Orders Full Investigation After Two Baby Deaths," ABC News, July 10, 2018, https://www.abc.net.au/news/2018-07-10/samoa-recalls-vaccines-order-investigation-after-baby-deaths/9971368; British Broadcasting Service, "How a Wrong Injection Helped Cause Samoa's Measles Epidemic," BBC News, December 2, 2019, https://www.bbc.com/news/world-asia-50625680.

181 **traveled to Samoa to meet with the prime minister:** Peter J. Hotez, Tasmiah Nuzhath, and Brian Colwell, "Combating Vaccine Hesitancy and Other 21st Century Social Determinants in the Global Fight Against Measles," *Current Opinion in Virology* 41 (2020): 1–7, doi: 10.1016/j.coviro.2020.01.001, PMID: 32113136; British Broadcasting Service, "How a Wrong Injection Helped Cause Samoa's Measles Epidemic"; Mehdi Hasan, "'Kids Died.' The Story of RFK Jr., Anti-Vaxxers, and a Measles Outbreak: Mehdi's Deep Dive,"

NOTES

MSNBC, July 6, 2023, https://www.msnbc.com/mehdi-on-msnbc/watch/-kids-died-the-story-of-rfk-jr-anti-vaxxers-and-a-measles-outbreak-mehdi-s-deep-dive-187033157936; James Robertson, "Charged Anti-Vaccine Activist Facing Up to Two Years," *Samoa Observer*, December 6, 2019, https://www.samoaobserver.ws/category/samoa/54398.

181 **most of which were in children under age five:** David Champredon et al., "Curbing the 2019 Samoa Measles Outbreak," *Lancet Infectious Dis*eases 20, no. 3 (2020): 287–88, doi: 10.1016/S1473-3099(20)30044-X, PMID: 32112762; United Nations, "Samoa Measles Outbreak Claims 70 Lives, Majority Are Children Under Five," UN News, December 10, 2019, https://news.un.org/en/story/2019/12/1053131; Julia Belluz, "Tiny Samoa Has Had Nearly 5,000 Measles Cases. Here's How It Got So Bad," Vox, December 18, 2019, https://www.vox.com/2019/12/18/21025920/measles-outbreak-2019-samoa.

182 **did not even have enough child-sized coffins:** Hasan, "'Kids Died.' The Story of RFK Jr., Anti-Vaxxers, and a Measles Outbreak: Mehdi's Deep Dive"; Miriam Berger, "Dozens of Samoan Children Have Died of Measles. So New Zealanders Are Sending Child-Size Coffins," *Washington Post*, December 7, 2019, https://www.washingtonpost.com/world/2019/12/07/dozens-samoan-kids-died-measles-so-new-zealanders-are-sending-child-sized-coffins-help/; Jonathan Barrett, "Decorated with Butterflies, Infant-Sized Coffins Sent to Measles-Ravaged Samoa," Reuters, December 8, 2019, https://www.reuters.com/article/us-health-measles-samoa/decorated-with-butterflies-infant-sized-coffins-sent-to-measles-ravaged-samoa-idUSKBN1YD01J.

182 **massive spike in measles activity:** World Health Organization, "Deaths from Democratic Republic of the Congo Measles Outbreak Top 6000," January 7, 2020, https://www.afro.who.int/news/deaths-democratic-republic-congo-measles-outbreak-top-6000.

182 **Ebola virus outbreak that had started in 2018:** World Health Organization, "Ebola Outbreak in the Democratic Republic of the Congo Declared a Public Health Emergency of International Concern," July 17, 2019, https://www.who.int/news/item/17-07-2019-ebola-outbreak-in-the-democratic-republic-of-the-congo-declared-a-public-health-emergency-of-international-concern.

182 **claimed far more lives:** United Nations, "Measles Claims More Than Twice as Many Lives Than Ebola in DR Congo," UN News, November 27, 2019, https://news.un.org/en/story/2019/11/1052321.

182 **a massive campaign:** World Health Organization, "Deaths from Democratic Republic of the Congo Measles Outbreak Top 6000."

182 **a nearly 50 percent increase since 2016:** World Health Organization, "Worldwide Measles Deaths Climb 50% from 2016 to 2019 Claiming over 207 500 Lives in 2019," November 12, 2020, https://www.who.int/news/item/12-11-2020-worldwide-measles-deaths-climb-50-from-2016-to-2019-claiming-over-207-500-lives-in-2019; Minal K. Patel et al., "Progress Toward Regional Measles Elimination—Worldwide, 2000–2019," *Morbidity and Mortality Weekly Report* 69, no. 45 (2020): 1700–705, doi: 10.15585/mmwr.mm6945a6, PMID: 33180759.

182 **dropped significantly over the past two decades:** Meredith G. Dixon et al., "Progress Toward Regional Measles Elimination—Worldwide, 2000–2020," *Morbidity and*

NOTES

Mortality Weekly Report 70, no. 45 (2021): 1563–69, doi: 10.15585/mmwr.mm7045a1, PMID: 34758014.

182 **threatened public health systems:** Jan Hoffman, "Measles Deaths Soared Worldwide Last Year, as Vaccine Rates Stalled," *New York Times,* November 12, 2020, https://www.nytimes.com/2020/11/12/health/measles-deaths-soared-worldwide-last-year-as-vaccine-rates-stalled.html.

CHAPTER 11: BOOSTER SHOTS

183 **"fighting on two fronts":** Bruce Schneier, "We Must Prepare for the Next Pandemic," *New York Times,* June 17, 2019, https://www.nytimes.com/2019/06/17/opinion/pandemic-fake-news.html.

184 **"fighting an infodemic":** "The COVID-19 Infodemic," *Lancet Infectious Diseases* 20, no. 8 (2020): 875, doi: 10.1016/S1473-3099(20)30565-X, PMID: 32687807.

185 **called it "China virus":** British Broadcasting Service, "Trump Says Coronavirus Worse 'Attack' Than Pearl Harbor," BBC News, May 7, 2020, https://www.bbc.com/news/world-us-canada-52568405.

185 **called it "Kung Flu":** British Broadcasting Service, President Trump Calls Coronavirus 'Kung Flu,'" BBC News, June 24, 2020, https://www.bbc.com/news/av/world-us-canada-53173436.

185 **widespread use of drugs like hydroxychloroquine:** Philip Bump, "The Rise and Fall of Trump's Obsession with Hydroxychloroquine," *Washington Post,* April 24, 2020, https://www.washingtonpost.com/politics/2020/04/24/rise-fall-trumps-obsession-with-hydroxychloroquine/.

185 **capitalize on this unique moment:** Tara Haelle, "This Is the Moment the Anti-Vaccine Movement Has Been Waiting For," *New York Times,* August 31, 2021, https://www.nytimes.com/2021/08/31/opinion/anti-vaccine-movement.html.

185 **joined forces with far-right extremists:** Nicholas Bogel-Burroughs, "Antivaccination Activists Are Growing Force at Virus Protests," *New York Times,* May 2, 2020, https://www.nytimes.com/2020/05/02/us/anti-vaxxers-coronavirus-protests.html.

186 **an ethical obligation (to protect them from disease):** Perri Klass and Adam J. Ratner, "Vaccinating Children Against Covid-19—The Lessons of Measles," *New England Journal of Medicine* 384, no. 7 (2021): 589–91, doi: 10.1056/NEJMp2034765, PMID: 33471977.

187 **falsehoods specifically to Black Americans:** Brandy Zadrozny and Char Adams, "Covid's Devastation of Black Community Used as 'Marketing' in New Anti-Vaccine Film," NBC News, March 11, 2021, https://www.nbcnews.com/news/nbcblk/covid-s-devastation-black-community-used-marketing-new-anti-vaxxer-n1260724.

187 **multiple anti-vaccine and anti-mandate rallies:** Keziah Weir, "How Robert F. Kennedy Jr. Became the Anti-Vaxxer Icon of America's Nightmares," *Vanity Fair,* May 13, 2021, https://www.vanityfair.com/news/2021/05/how-robert-f-kennedy-jr-became-anti-vaxxer-icon-nightmare; Sarah Fortinsky and Aileen Graef, "Robert F. Kennedy Jr. Invokes Nazi Germany in Offensive Anti-Vaccine Speech," CNN, January 24, 2022, https://www.cnn

NOTES

.com/2022/01/23/politics/robert-f-kennedy-nazi-germany-offensive-anti-vaccine-speech/index.html.

187 **promotes the false thesis:** Adam Nagourney, "A Kennedy's Crusade Against Covid Vaccines Anguishes Family and Friends," *New York Times*, February 26, 2022, https://www.nytimes.com/2022/02/26/us/robert-kennedy-covid-vaccine.html.

187 **uptake of COVID-19 vaccines has been less than optimal:** Centers for Disease Control and Prevention (CDC), "COVID-19 Vaccinations in the United States," https://covid.cdc.gov/covid-data-tracker/#vaccinations_vacc-total-admin-rate-total.

188 **routine childhood vaccination:** Alex Shephard, "The Right's War on Vaccines Is Spreading Beyond Covid," *The New Republic*, April 20, 2022, https://newrepublic.com/article/166140/anti-vaccine-spreading-beyond-covid; Eric Lutz, "The Right's War on COVID Vaccine Mandates Is About to Get Scary," *Vanity Fair*, September 14, 2021, https://www.vanityfair.com/news/2021/09/the-rights-war-on-covid-vaccine-mandates-is-about-to-get-scary; Haelle, "This Is the Moment the Anti-Vaccine Movement Has Been Waiting For."

188 **pushing back against all mandates:** Shephard, "The Right's War on Vaccines Is Spreading Beyond Covid."

188 **wrote the foreword:** Children's Health Defense, *Measles Book: Thirty-Five Secrets the Government and the Media Aren't Telling You About Measles and the Measles Vaccine*, foreword by Robert F. Kennedy Jr. (New York: Skyhorse Publishing, 2021).

188 **significant increase in resistance:** Lunna Lopes et al., "KFF COVID-19 Vaccine Monitor: December 2022," December 16, 2022, https://www.kff.org/coronavirus-covid-19/poll-finding/kff-covid-19-vaccine-monitor-december-2022/.

188 **refusing rabies vaccination for their dogs:** Matt Motta, Gabriella Motta, and Dominik Stecula, "Sick as a Dog? The Prevalence, Politicization, and Health Policy Consequences of Canine Vaccine Hesitancy (CVH)," *Vaccine* 41, no. 41 (2023): 5946–50, doi: 10.1016/j.vaccine.2023.08.059, PMID: 37640567.

188 **a "trust problem":** Ed Pertwee, Clarissa Simas, and Heidi J. Larson, "An Epidemic of Uncertainty: Rumors, Conspiracy Theories and Vaccine Hesitancy," *Nature Medicine* 28, no. 3 (2022): 456–59, doi: 10.1038/s41591-022-01728-z, PMID: 35273403.

188 **driving forces in modern hesitancy:** Pertwee, Simas, and Larson, "An Epidemic of Uncertainty: Rumors, Conspiracy Theories and Vaccine Hesitancy"; Majdi M. Sabahelzain, Kenneth Hartigan-Go, and Heidi J. Larson, "The Politics of Covid-19 Vaccine Confidence," *Current Opinion in Immunology* 71 (2021): 92–96, doi: 10.1016/j.coi.2021.06.007, PMID: 34237648.

189 **Political affiliation has become a medical risk factor:** Jacob Wallace, Paul Goldsmith-Pinkham, and Jason L. Schwartz, "Excess Death Rates for Republican and Democratic Registered Voters in Florida and Ohio During the COVID-19 Pandemic," *JAMA Internal Medicine* 183, no. 9 (2023): 916–23, doi: 10.1001/jamainternmed.2023.1154, PMID: 37486680.

189 **hindered COVID-19 vaccination uptake:** Moises Velasquez-Manoff, "The Anti-Vaccine Movement's New Frontier," *New York Times*, May 25, 2022, https://www.nytimes.com/2022/05/25/magazine/anti-vaccine-movement.html.

NOTES

189 **overturning a 1979 state supreme court decision:** Supreme Court of Mississippi, *Brown v. Stone* (No. 51553), https://law.justia.com/cases/mississippi/supreme-court/1979/51553-0.html; Dorit R. Reiss and Arthur L. Caplan, "A Judge's Infuriating Ruling on Vaccination Puts Mississippi's Children at Risk," STAT News, April 27, 2023, https://www.statnews.com/2023/04/27/religious-exemption-vaccines-children-mississippi/.

189 **a lawsuit funded by the Texas anti-vaccination group:** Will Stribling, "Mississippi Now Allows Religious Exemptions for Childhood Vaccinations," Mississippi Public Broadcasting (MBP) News, July 18, 2023, https://www.mpbonline.org/blogs/news/mississippi-now-allows-religious-exemptions-for-childhood-vaccinations-some-pediatricians-fear-a-re/; Reiss and Caplan, "A Judge's Infuriating Ruling on Vaccination Puts Mississippi's Children at Risk."

189 **inhibited distribution of routine childhood vaccines:** Amy G. Feldman, Sean T. O'Leary, and Lara Danziger-Isakov, "The Risk of Resurgence in Vaccine-Preventable Infections Due to Coronavirus Disease 2019–Related Gaps in Immunization," *Clinical Infectious Diseases* 73, no. 10 (2021): 1920–23, doi: 10.1093/cid/ciab127, PMID: 33580243.

189 **dropped precipitously:** Jeanne M. Santoli et al., "Effects of the COVID-19 Pandemic on Routine Pediatric Vaccine Ordering and Administration—United States, 2020," *Morbidity and Mortality Weekly Report* 69, no. 19 (2020): 591–93, doi: 10.15585/mmwr.mm6919e2, PMID: 32407298.

190 **fewer than half of Michigan infants:** Cristi A. Bramer et al., "Decline in Child Vaccination Coverage During the COVID-19 Pandemic—Michigan Care Improvement Registry, May 2016–May 2020," *Morbidity and Mortality Weekly Report* 69, no. 20 (2020): 630–31, doi: 10.15585/mmwr.mm6920e1, PMID: 32437340.

190 **fell by about ten percentage points:** Sara M. Bode et al., "COVID-19 and Primary Measles Vaccination Rates in a Large Primary Care Network," *Pediatrics* 147, no. 1 (2021): e2020035576, doi: 10.1542/peds.2020-035576, PMID: 33214332.

190 **Similar patterns emerged:** Tasmia Nuzhath et al., "Childhood Immunization During the COVID-19 Pandemic in Texas," *Vaccine* 39, no. 25 (2021): 3333–37, doi: 10.1016/j.vaccine.2021.04.050, PMID: 34020814.

190 **back to their pre-pandemic levels:** Sean T. O'Leary et al., "US Primary Care Providers' Experiences and Practices Related to Routine Pediatric Vaccination During the COVID-19 Pandemic," *Academic Pediatrics* 22, no. 4 (2022): 559–63, doi: 10.1016/j.acap.2021.10.005, PMID: 34757024.

190 **well below the required 95 percent threshold:** Ranee Seither et al., "Vaccination Coverage with Selected Vaccines and Exemption Rates Among Children in Kindergarten—United States, 2020–21 School Year," *Morbidity and Mortality Weekly Report* 71, no. 16 (2022): 561–68, doi: 10.15585/mmwr.mm7116a1, PMID: 35446828.

190 **may be long-lasting:** Benjamin Mueller and Jan Hoffman, "Routine Childhood Vaccinations in the U.S. Slipped During the Pandemic," *New York Times*, April 21, 2022, https://www.nytimes.com/2022/04/21/health/pandemic-childhood-vaccines.html; Douglas J. Opel et al., "The Legacy of the COVID-19 Pandemic for Childhood

NOTES

Vaccination in the USA," *Lancet* 401, no. 10370 (2023): 75–78, doi: 10.1016/S0140-6736(22)01693-2, PMID: 36309017.

190 **More than 40 percent of those affected:** Elizabeth C. Tiller et al., "Notes from the Field: Measles Outbreak—Central Ohio, 2022–2023," *Morbidity and Mortality Weekly Report* 72, no. 31 (2023): 847–49, doi: 10.15585/mmwr.mm7231a3, PMID: 37535476.

190 **his infection was acquired in the United States:** Ruth Link-Gelles et al., "Public Health Response to a Case of Paralytic Poliomyelitis in an Unvaccinated Person and Detection of Poliovirus in Wastewater—New York, June–August 2022," *Morbidity and Mortality Weekly Report* 71, no. 33 (2022): 1065–68, doi: 10.15585/mmwr.mm7133e2, PMID: 35980868; Marie E. Wang and Adam J. Ratner, "Clinical Progress Note: Poliomyelitis," *Journal of Hospital Medicine* 18, no. 1 (2023): 61–64, doi: 10.1002/jhm.12989, PMID: 36314273.

191 **MMR coverage in the United Kingdom:** Tony Kirby, "MMR Vaccination in England Falls Below Critical Threshold," *Lancet Infectious Diseases* 22, no. 4 (2022): 453, doi: 10.1016/S1473-3099(22)00169-4, PMID: 35338872; Helen Bedford and Helen Donovan, "We Need to Increase MMR Uptake Urgently," *British Medical Journal* 376 (2022): o818, doi: 10.1136/bmj.o818, PMID: 35354576.

191 **slow and incomplete:** World Health Organization, "COVID-19 Pandemic Leads to Major Backsliding on Childhood Vaccinations, New WHO, UNICEF Data Shows," July 15, 2021, https://www.who.int/news/item/15-07-2021-covid-19-pandemic-leads-to-major-backsliding-on-childhood-vaccinations-new-who-unicef-data-shows.

191 **more than 23 million children:** World Health Organization, "COVID-19 Pandemic Leads to Major Backsliding on Childhood Vaccinations, New WHO, UNICEF Data Shows."

191 **Global MCV1 coverage dropped:** Anna A. Minta et al., "Progress Toward Regional Measles Elimination—Worldwide, 2000–2021," *Morbidity and Mortality Weekly Report* 71, no. 47 (2022): 1489–95, doi: 10.15585/mmwr.mm7147a1, PMID: 36417303.

192 **children missed out on vitamin A supplementation:** Andreas Hasman et al., "COVID-19 Caused Significant Declines in Regular Vitamin A Supplementation for Young Children in 2020: What Is Next?" *BMJ Global Health* 6, no. 11 (2021): e007507, doi: 10.1136/bmjgh-2021-007507, PMID: 34785507.

192 **substantially higher risk of measles complications:** Gregory D. Hussey and Max Klein, "A Randomized, Controlled Trial of Vitamin A in Children with Severe Measles," *New England Journal of Medicine* 323, no. 3 (1990): 160–64, doi: 10.1056/NEJM199007193230304, PMID: 2194128.

192 **"the eyes of the hippopotamus":** Nick Paumgarten, "The Message of Measles," *The New Yorker*, September 2, 2019, https://www.newyorker.com/magazine/2019/09/02/the-message-of-measles.

192 **we just don't know by how much:** Meredith G. Dixon et al., "Progress Toward Regional Measles Elimination—Worldwide, 2000–2020," *Morbidity and Mortality Weekly Report* 70, no. 45 (2021): 1563–69, doi: 10.15585/mmwr.mm7045a1, PMID: 34758014.

NOTES

192 **a "perfect storm of conditions":** World Health Organization, "UNICEF and WHO Warn of Perfect Storm of Conditions for Measles Outbreaks, Affecting Children," April 27, 2022, https://www.who.int/news/item/27-04-2022-unicef-and-who-warn-of-perfect-storm-of-conditions-for-measles-outbreaks-affecting-children.

193 **thousands of cases in the mid-pandemic period:** World Health Organization, "UNICEF and WHO Warn of Perfect Storm of Conditions for Measles Outbreaks, Affecting Children."

193 **massive outbreak driven by vaccine hesitancy:** Tendai Marima and Stephanie Nolen, "More Than 700 Children Have Died in a Measles Outbreak in Zimbabwe," *New York Times*, September 24, 2022, https://www.nytimes.com/2022/09/24/health/measles-outbreak-zimbabwe.html.

193 **lost seven children to measles:** Evidence Chenjerai, "These Zimbabwean Churches Disavowed Vaccines. Then Measles Came for Their Kids," *Global Press Journal*, March 8, 2023, https://globalpressjournal.com/africa/zimbabwe/faith-prohibited-vaccinations-measles-killed-children/.

196 **"What We Don't See":** Margaret Kendrick Hostetter, "What We Don't See," *New England Journal of Medicine* 366, no. 14 (2012): 1328–34, doi: 10.1056/NEJMra1111421, PMID: 22475596.

196 **changed childhood from a perilous period:** Perri Klass, *The Best Medicine: How Science and Public Health Gave Children a Future* (New York: Norton, 2022).

198 **"pre-bunking" in addition to debunking:** Ed Pertwee, Clarissa Simas, and Heidi J. Larson, "An Epidemic of Uncertainty: Rumors, Conspiracy Theories and Vaccine Hesitancy," *Nature Medicine* 28, no. 3 (2022): 456–59, doi: 10.1038/s41591-022-01728-z, PMID: 35273403.

198 **pockets of nonmedical exemptions:** Jacqueline K. Olive et al., "The State of the Antivaccine Movement in the United States: A Focused Examination of Nonmedical Exemptions in States and Counties," *PLoS Medicine* 15, no. 6 (2018): e1002578, doi: 10.1371/journal.pmed.1002578, PMID: 29894470.

198 **kilometers rather than counties:** Nina B. Masters et al., "Fine-Scale Spatial Clustering of Measles Nonvaccination That Increases Outbreak Potential Is Obscured by Aggregated Reporting Data," *Proceedings of the National Academies of Science of the USA* 117, no. 45 (2020):28506–14, doi: 10.1073/pnas.2011529117, PMID: 33106403.

198 **we miss important information:** Local Burden of Disease Vaccine Coverage Collaborators, "Mapping Routine Measles Vaccination in Low- and Middle-Income Countries," *Nature* 589, no. 7842 (2021): 415–19, doi: 10.1038/s41586-020-03043-4, PMID: 33328634.

198 **identify clusters of vaccine misinformation online:** Vaccine Confidence Project: https://www.vaccineconfidence.org/; Project VCTR: https://projectvctr.com/; Stanford Internet Observatory: https://cyber.fsi.stanford.edu/io.

199 **long-term aid and local capacity building:** David N. Durrheim, "Measles Elimination—Using Outbreaks to Identify and Close Immunity Gaps," *New England Journal of Medicine* 375, no. 14 (2016): 1392–93, doi: 10.1056/NEJMe1610620, PMID: 27705259.

NOTES

200 **"measles can and should be eradicated":** World Health Organization, "Proceedings of the Global Technical Consultation to Assess the Feasibility of Measles Eradication, 28–30 July 2010," *Journal of Infectious Diseases* 204, suppl. 1 (2011): S4–S13, doi: 10.1093/infdis/jir100, PMID: 21666191.

200 **eradicated after a massive international campaign:** Amanda Kay McVety, *The Rinderpest Campaigns: A Virus, Its Vaccines, and Global Development in the Twentieth Century* (New York: Cambridge University Press, 2018); Clive A. Spinage, *Cattle Plague* (New York: Kluwer Academic/Plenum Publishers, 2003).

201 **a particularly fun and approachable episode:** Moo Cows, Moo Problems, *This Podcast Will Kill You*, episode 52, June 23, 2020, https://thispodcastwillkillyou.com/2020/06/23/episode-52-rinderpest-moo-cows-moo-problems.

201 **a savings of $58 for every $1 spent:** Sachiko Ozawa et al., "Return on Investment from Childhood Immunization in Low- and Middle-Income Countries, 2011–20," *Health Affairs* (Millwood) 35, no. 2 (2016): 199–207, doi: 10.1377/hlthaff.2015.1086, PMID: 26858370.

201 **aid eradication logistics:** Jeffrey C. Mariner et al., "Rinderpest Eradication: Appropriate Technology and Social Innovations," *Science* 337, no. 6100 (2012): 1309–12, doi: 10.1126/science.1223805, PMID: 22984063.

202 **dealing with inequities in vaccine availability:** Amy K. Winter et al., "Feasibility of Measles and Rubella Vaccination Programmes for Disease Elimination: A Modelling Study," *Lancet Global Health* 10, no. 10 (2022): e1412–e1422, doi: 10.1016/S2214-109X(22)00335-7, PMID: 36113527; Pratima L. Raghunathan and Walter Orenstein, "Investing in Global Measles and Rubella Elimination Is Needed to Avert Deaths and Advance Health Equity," *Lancet Global Health* 10, no. 10 (2022): e1363–e1364, doi: 10.1016/S2214-109X(22)00388-6, PMID: 36113510.

202 **"it all depends on whose baby has the measles":** Franklin Delano Roosevelt, "Address on Receiving the 1935 Award for Distinguished Service to Agriculture, Chicago, Illinois," December 9, 1935, https://www.presidency.ucsb.edu/documents/address-receiving-the-1935-award-for-distinguished-service-agriculture-chicago-illinois; Franklin Delano Roosevelt, campaign speech. October 24, 1928, http://www.fdrlibrary.marist.edu/_resources/images/msf/msf00285. Eleanor Roosevelt used a similar adage: Eleanor Roosevelt, "My Day, June 20, 1958," *The Eleanor Roosevelt Papers Digital Edition*, https://www2.gwu.edu/~erpapers/myday/displaydoc.cfm?_y=1958&_f=md004150.

Index

AAP (American Academy of Pediatrics), 102, 118
aboriginal peoples, 48–49
ACIP (Advisory Committee on Immunization Practices), 102
Affordable Care Act (ACA), 110–11
Afghanistan, 192–93
African nations
  Ebola outbreaks, 7, 56
  slave trade, 41, 43, 73–74
  smallpox inoculation practice, 73–74
  vaccination campaigns, 96, 103, 105–6, 139
Age of Autism website, 155–56
Alabama, 108
Alaska, 56–57, 108
All Your Vaccine Questions Answered (PEACH publication), 171
American Academy of Pediatrics (AAP), 102, 118
amnesia (immune), xviii–xix, xxv–xxvi, 143–47, 181, 194–97
anticipatory guidance, 197–98
anti-vaccine sentiment, 145–64, 168–80
  author's encounters with, xx–xxii
  autism misinformation, 151–57, 163, 168, 171, 174, 178
  background, 141, 145–47
  of COVID pandemic, 98, 118–19, 128, 185–90
  national databases as tools for, 128

  organizations dedicated to, 126–27, 150, 159–60
  public outreach to shift, 174–75, 177–79
  public trust issues, 97–98, 148–50, 169–80
  religion's role in, 148–49, 154, 168–80, 193
  school mandate exemptions, 158–63, 172, 173–74, 179
  of Trump, 163–64, 174
Arkansas, 109–11, 116–17
Australia, 48–49
autism, 151–57, 163, 168, 171, 174, 178

Bad Faith (Offit), 148
The Best Medicine (Klass), 196
Biel, Jessica, 162
Black, Francis, 92
Bloomberg, Michael, 169
Boston, 44–46, 73–74
Bradley University, 133–34
Brevig Mission, 57–59
Brooklyn, 167–70, 172–81, 189
Brown, Jerry, 161
bubonic plague, 55–56
Bumpers, Betty, 116–17
Bumpers, Dale, 116–17
Burton, Dan, 153
Butler, William, 62–63, 66

Cakobau (king), 48–50
Califano, Joseph, Jr., 117–18, 119
California, 117, 157–63, 185, 198

INDEX

Callous Disregard (McCarthy), 153
Calvignac-Spencer, Sébastien, 60, 61
Carrey, Jim, 156
Carter, Jimmy, 115–20, 121, 122
Carter, Rosalynn, 116–17
cattle plague, 200–201
CDC (Centers for Disease Control and Prevention)
  COVID-19 recommendations, 184
  on elimination of measles, 104, 106, 107, 119, 140
  on health disparities, 149–50
  immunization policies and programs, 102, 137, 138
  on live vs. killed vaccine, 96
  name change, 94
  public outreach programs, 114, 118
  on religious anti-vaccination sentiments, 154
  reporting system, 127–28
  on school mandates, 158–59
  Wakefield on, 163
Celebrezze, Anthony, Sr., 95
Centers for Disease Control and Prevention. *See* CDC
Central African Republic, 139
Cherry, James, 158
Chicago, 62, 133–34
Children's Health Defense, 187, 188
China, 11, 73, 185
cholera, 21, 34, 35–36
Clark County (Washington), 174
Clinton, Bill, 136–38, 146, 148–49
Coleman, Joshua, 185, 187–88
college measles outbreaks, 129–34
Collingwood, Charles, 95–96
colonialism and commerce, 37–51
  background, 37–39
  endemicity and, 39–40
  epidemiology's birth and, 29–30, 36
  non-indigenous population outbreaks, 45–47, 62
  prevention options spread through, 73–74
  virgin soil theory and depopulation, 40–45, 47–51
Communicable Diseases Center. *See* CDC

Comprehensive Child Immunization Act, 136–38
Conis, Elena, 119
Conrad, J. L., 107
Cook, David, 43
COVID-19 pandemic
  anti-vaccine sentiment during, 98, 118–19, 128, 185–90
  data-driven approach to, 11–12, 15, 35
  disruption of health services during, 189–90, 191–93
  early days of, 14–15, 37–39, 184–85
  lessons to be learned from, 195–98
  mis- and disinformation about, 128, 183–88
  social determinants of health and, 53–54, 69–70
  vaccine for, 11, 100, 132
  virus, 4–5, 7, 11, 15
Cruickshank, John, 48–49
Cutter Laboratories, 124–25

Day, Ed, 175
de Blasio, Bill, 170, 175, 177
Deer, Brian, 151
Democratic Republic of the Congo, 139, 182
Denmark. *See* Faroe Islands
Diamond, Jared, 40
diphtheria, tetanus, and pertussis (DTP) vaccines, 125–27, 137
Disneyland measles outbreak, 158–60, 198
Dissatisfied Parents Together, 126–27, 150, 159, 172
The Doctor Who Fooled the World (Deer), 151
Downs, Jim, 36
Drinkwater, Harry, 64–65
DTP (diphtheria, tetanus, and pertussis) vaccines, 125–27, 137
Dying of Whiteness (Metzl), 110

Ebola outbreaks, 7, 56, 182
Edmonston, David, 85, 86–88, 90, 92
Eisenhower, Dwight D., 97, 101
Eisenstein, Mayer, 171

INDEX

elimination and eradication goals, 100–101, 103–11, 116, 119–21, 138–41, 199–202. *See also* mass immunization
Enders, John, 81–88, 90–92, 95–96, 103, 140
Engaging in Medical Education with Sensitivity (EMES) Initiative, 177–79, 180, 189, 195, 197–98
England, 44–51. *See also* United Kingdom
epidemiology, 7–8, 16, 26–32, 35–36, 110–11, 132–36
eradication and elimination goals, 100–101, 103–11, 116, 119–21, 138–41, 199–202
Ethiopia, 192–93
eugenics movement, 90
European conquests. *See* colonialism and commerce

Fannin, Shirley, 114, 146–47
Faroe Islands
 COVID-19 pandemic, 35
 lessons learned from, 34–36
 measles epidemic, 17–20, 22–29, 31–36
 social determinants of health in, 28–30
Fauci, Anthony, 187
FDA (Food and Drug Administration), 127–28
Fiji, 48–51
Fiscus, Michelle, 186
Fisher, Barbara Loe, 150, 159–60, 172
Flake, Emily, 160
Fletcher, Jackie, 150
Foege, William, 105, 140–41
Food and Drug Administration (FDA), 127–28
Fort Lewis College, 131–32, 134
Fracastoro, Girolamo, 21
France, 105
Francis, Thomas, Jr., 84
Friedman, Moshe, 176

Geist, Otto, 57
Generation Rescue, 156
"German measles," 107–8. *See also* MMR vaccine
Germany, 176

Ghebreyesus, Tedros Adhanom, 184
Glasgow, 66–67, 68
Global Smallpox Eradication Program, 105
Gottsdanker v. Cutter Laboratories, 124–25
"Green Our Vaccines" rally, 156
Guns, Germs, and Steel (Diamond), 40

Haelle, Tara, 185
Halliday, James, 66–67
Handler, Hillel, 176–77
Handley, J. B., 155–56
Hawaii, 51
health access inequities, 103–4, 110–11, 129, 134–38, 188–89, 195–97
Henle, Jacob, 21–22
hepatitis, 89, 123–24, 149
Herrman, Charles, 77–79
hesitancy to vaccinate. *See* anti-vaccine sentiment
Hilleman, Maurice, 88, 89, 92, 93, 95–96, 107
Hinman, Alan, 152
HIV (human immunodeficiency virus), 5, 59
Holloway, Ann, 92
Holm, Christen Severin, 17–18, 25
Holmes, Oliver Wendell, Sr., 22
Home, Francis, 75–77, 81
Hostetter, Margaret, 196
How the Other Half Lives (Riis), xxv, 67
Hultin, Johan, 57–59
human immunodeficiency virus (HIV), 5, 59

immigrants, 53–54, 155–57
immune amnesia, xviii–xix, xxv–xxvi, 143–47, 181, 194–97
Inca civilization, 43
India, 61–62, 73
Indiana, 153–55
indigenous people of "New World," 40–44
influenza, 4, 7, 40, 56–57, 98
Informed Consent Action Network, 189
Ingalls, Thomas, 84–85
innate (or immune) cells, 143–45
inoculation, 72–79
Israel, 173

INDEX

JABS (Justice, Awareness, and Basic Support), 150, 151
Jacobson v. Massachusetts (1905), 174
Jewish communities. *See* Orthodox Jewish communities
Johnson, Lyndon B., 104
Josefa, Ratu, 49
Justice, Awareness, and Basic Support (JABS), 150, 151

Katz, Samuel, 90–91, 92
Kellogg's campaign, 118
Kempe, C. Henry, 91, 92
Kennedy, John F., 94, 96, 100–101
Kennedy, Robert F., Jr., 156, 161, 162, 163–64, 181, 187, 188
Kent State University, 129–30, 134
Kerala, 61–62
Kitab Al-Jadari Wal-Hasba (*Treatise on Smallpox and Measles*), 23
Klass, Perri, 186, 196
Koch, Robert, 36
Koplik spots, 5
Krugman, Saul, 92, 123–24, 150

Langmuir, Alexander, 94
Larson, Heidi, 152, 188
Lindenberger, Ethan, 174
Li Wenliang, 12
London, 35, 62–65
Los Angeles County, 114–15, 146–47

Magnalia Christi Americana (Mather), 44–45
Manicus, August, 19–20, 22–25, 27–28, 30, 32–34, 77
Marcus, Blima, 177–78, 179, 195
mass immunization, 99–111, 113–41
  current obstacles to, 99–100, 113–14, 134–35, 147–64. *See also* anti-vaccine sentiment
  elimination as public health goal, 100–101, 103–11, 116, 119–21, 138–41, 199–202
  future directions for, 193–202
  historical context, 96–98, 102–3, 129–34
  primary vaccine failures, 121

public outreach programs, 116–19, 136–39
public trust and, 100, 102, 113, 120–28, 148–51, 169–80
school mandates, 109–10, 114–15, 118–20, 130–33, 150, 158–63
Mather, Cotton, 44–46, 73–74
Mayan civilization, 43
McCarthy, Jenny, 153, 156
McNeill, William H., 39
measles
  absence of repeat infections, 32–33
  childhood fatalities, 63–68
  contagious period, 32
  early case descriptions, 16
  elimination and eradication goals. *See* mass immunization
  emergence of, 8, 61–62
  endemicity and urbanization, 39–40, 61–64, 66
  incubation period, 25–27, 31–32, 33, 42–43
  outbreaks. *See* measles outbreaks
  prevention. *See* measles prevention
  quarantine effectiveness, 32–33
  reproduction number, 7–8
  spread of, 6–8, 32, 62–70. *See also* measles outbreaks
  storytelling and, 12–16
  symptoms and diagnosis, 5–6
  virus, 3–8, 54–56, 59–62, 143–46
Measles Book (Children's Health Defense), 188
measles elimination. *See* mass immunization
measles outbreaks (major). *See also* colonialism and commerce
  author's personal story, xviii–xxv
  in California, 114–15, 158–60, 198
  in colleges, 129–34
  Faroe Islands, 17–20, 22–29, 31–36
  in religious communities, 147–48, 153–55, 166, 168–69, 172–81
  in Somali American community, 155–57
  in Texarkana, 109–11
  in Ukraine, 165–66, 172–73, 174
  in United Kingdom, 105, 150–52, 167, 191
  vaccinated population outbreaks, 131–33
"measles parties," 80–81

258

INDEX

measles prevention, 71–98
  author's personal story, xi–xviii
  elimination and eradication goals. *See* mass immunization
  future directions for, 193–202
  historical attempts at, 72–80
  reproduction number, 7–8
  vaccination vs. natural immunity, 71–72
  vaccine approval, 93–96
  vaccine combination (MMR), 108, 121, 132–33, 150–53, 155–57, 167–69, 181, 190–91
  vaccine development, 80–88
  vaccine rollout, 96–98, 102–3
  vaccine trials, 88–92
Medicaid coverage, 110–11, 137
Medical Facts and Experiments (Home), 75–77
Medical Racism: The New Apartheid (film), 187
memory (immunological), 143–45
Merck, 88, 89, 92, 93, 95, 103, 107–8
metzitzah b'peh, 169–70
Metzl, Jonathan M., 110
miasma theory, 21–22, 24, 33–34, 36
Michigan, 190
Minnesota, 155–57
MIS-C (multisystem inflammatory syndrome in children), 186–87
Mission, Measles: The Story of a Vaccine (film), 103
Mississippi, 189
MMR vaccine, 108, 121, 132–33, 150–53, 155–57, 167–69, 181, 190–91
Montagu, Edward Wortley, 74
Montagu, Mary Wortley, 74
morbilliviruses, 54–56
multisystem inflammatory syndrome in children (MIS-C), 186–87
mumps, 166–69. *See also* MMR vaccine
Murray, Roderick, 95

National Childhood Vaccine Injury Act of 1986 (NCVIA), 127–28
National Federation of Independent Business v. Sebelius (2012), 110
National Institutes of Health (NIH), 93–94, 95
National Vaccine Advisory Committee (NVAC), 127, 135–36, 146
National Vaccine Information Center (NVIC), 159–60, 171
National Vaccine Injury Compensation Program (NVICP), 127, 128–29
National Vaccine Program, 127
NCVIA (National Childhood Vaccine Injury Act of 1986), 127–28
New England First Fruits, 44
New York (state), 77–78, 134, 163–64, 179, 190–91
New York City, xix–xx, 14–15, 53–54, 62, 67, 166–81, 189, 190–91
Newsom, Gavin, 162
Nigeria, 192–93
NIH (National Institutes of Health), 93–94, 95
Nipah virus, 61–62
Nixon, Richard, 107
Nolsø, Mr., 18–19
NVAC (National Vaccine Advisory Committee), 127, 135–36, 146
NVIC (National Vaccine Information Center), 159–60, 171
NVICP (National Vaccine Injury Compensation Program), 127, 128–29

Obamacare. *See* Affordable Care Act
Observationes Medicae (Sydenham), 23
Offit, Paul, 89, 124–25, 148, 151
Ohio, 118, 129–30, 157, 190
Onesimus (enslaved person), 73–74
Oregon, 108
Orenstein, Walter, 99, 147, 152
Orthodox Jewish communities
  anti-vaccine sentiment within, 168–80
  measles outbreaks, 166, 168–69, 172–81
  mumps outbreaks, 167–68
Orthodox Jewish Nurses Association, 177

Pacific Island nations, 47–51
Palevsky, Lawrence, 176–77
Pan, Richard, 161, 162

INDEX

Pan American Health Organization (PAHO), 140
Panum, Peter Ludvig, 19–20, 22–35, 77, 195
paralytic polio, 125, 190–91
paramyxoviruses, 54–55, 61–62
Parents Educating and Advocating for Children's Health (PEACH), 171–72, 175, 177, 178
Parents Informed & Educated (*PIE*; EMES publication), 178
Pasteur, Louis, 36
Patient Protection and Affordable Care Act (ACA), 110–11
Peanuts cartoons, 106
Peebles, Thomas, 84–86, 95–96
peste des petits ruminants virus, 55
Philadelphia, 62, 147–48
Philippines, 174
Pirquet, Clemens von, 145
plagues, 55–56, 200–201
Plagues and Peoples (McNeill), 39
polio and polio vaccine
  adverse effects, 125
  development and trials, 82–84, 86–87, 96–97, 124–25
  eradication efforts, 82, 97, 100–101, 140–41, 190–91, 194–202
  virus, 7
Poliomyelitis Vaccination Assistance Act, 101
prevention. *See* measles prevention
puerperal fever, 22

R_0 ("R-naught"), 7–8
racism
  anti-vaccine sentiment and, 187
  in biomedical research, 122–23, 149–50
  slavery, 39–40, 41, 43, 73–74
  social determinants of health and, 68, 69–70, 134–35, 195–97
al-Razi, Abu Bakr Muhammad ibn Zakariya, 23
The Real Anthony Fauci (Kennedy), 187
Regenburg, Carl, 17–19
regressive neurological condition, 151
religion's role in measles outbreaks, 148–49, 154, 168–81, 193

reproduction number (R_0), 7–8
Reyes, Anita, 125
Rhazes, 23, 60
Riis, Jacob, xxv, 67
rinderpest virus, 200–201
RNA virus genetics, 54–61, 143–44
Robbins, Frederick, 82–84, 92
Robinson, Hercules, 48
Rockland County (New York), 166–68, 172–81, 190–91
Romania, 153–54
Roosevelt, Franklin Delano, 202
rotavirus, 9–10, 144
rubella, 107–8. *See also* MMR vaccine
Rucker, William Colby, 80

Sabin, Albert, 82, 83–84, 87
Salk, Jonas, 82, 83–84, 87
Samoa, 181–82
SARS-CoV-2 virus, 4–5, 7, 11, 15. *See also* COVID-19 pandemic
SARS outbreak (2002–2004), 12
Satcher, David, 149–50
Schaffer, Amanda, 170–71
Schneider, Rob, 162
Schneier, Bruce, 183–84
school vaccine mandates
  California legislature on, 161–63
  on COVID-19 vaccines, 189
  elimination goal and, 150
  exemptions, 158–63, 172, 173–74, 179
  mass immunization strategy and, 109–10, 114–15, 118–20, 130–33, 150, 158–63
  usefulness of, 109–10, 114–15
Scientology, Church of, 176
Scotland, 17–20, 28–36, 66–67, 68
Semmelweiss, Ignaz, 22
Sencer, David, 104
Shephard, Alex, 188
Silber, Chany, 170–71
SLAM (measles virus receptor), 4, 5, 6, 143, 144
slavery, 39–40, 41, 43, 73–74
A Slice of PIE (EMES publication), 178
smallpox
  colonialism and, 40, 41–45, 49

260

INDEX

eradication strategies and results, 99, 100, 105–6, 200
historical context, 21, 23, 31, 33, 60
measles comparison, 140
prevention through inoculation, 72–76
reproduction number, 7
Smith, Tara C., 129–30
Snow, John, 35–36
social determinants of health, xxiii–xxiv, 28–30, 36, 53–54, 64–70, 134–35, 149–50, 195–97
Somali American community, 155–57
Somalia, 192–93
Spanish colonies, 42–44
St. Louis, 62
Stokes, Joseph, Jr., 93
storytelling in medicine
background, 9–11
case presentations, 11–16
data-driven approach, 11
Sydenham, Thomas, 23
syphilis, 40, 122–23, 149–50

The Taming of a Virus (TV special), 95–96
Taubenberger, Jeffery, 58–59
Teller Mission, 57–59
Tenpenny, Sherri, 171
Terry, Luther, 94, 95, 96
Texarkana, 109–11
Texas, 109–11, 190
Thomas, Stephen B., 149
Timonius, Emanuel, 73–74
Toledo, Viceroy Francisco de, 44
Treatise on Smallpox and Measles (*Kitab Al-Jadari Wal-Hasba*), 23
Trump, Donald, 163–64, 174, 185
Tuskegee experiments, 122–23, 149–50

Ukraine, 165–66, 172–73, 174
United Kingdom, 105, 150–52, 167, 191. *See also* England
university measles outbreaks, 129–34
urbanization, 39–40, 61–64, 66

Vaccinated (Offit), 89
vaccinated population outbreaks, 131–33

Vaccination Assistance Act of (VAA), 102, 103–4
Vaccine Adverse Event Reporting System (VAERS), 127–28
Vaccine Confidence Project, 152, 188
vaccine hesitancy. *See* anti-vaccine sentiment
vaccines. *See* COVID-19 pandemic; mass immunization; measles prevention; polio and polio vaccine
The Vaccine Safety Handbook (PEACH), 171–72, 177, 178
Vaccines for Children (VFC), 137–38, 189–90
VAERS (Vaccine Adverse Event Reporting System), 127–28
value of R, 179
variolation, 72–77
VFC (Vaccines for Children), 137–38, 189–90
viral propagation research, 81–83
Virchow, Rudolf, 34, 60
virgin soil theory, 40–42, 47–50
vitamin A deficiency, 65–66, 192

Wakefield, Andrew, 150–57, 163–64, 177
Waldo, Dr., 65
Waldorf schools, 159, 173–74
Walsh, Dr., 65
Washington (state), 174
Weller, Thomas, 82–84
"What We Don't See" (Hostetter), 196
Willowbrook State School experiments, 123–24, 150
Wilson, John, 125–26
Witherill, Liston, 114
World Health Organization (WHO), 105, 139, 172, 184, 191, 200
Wyoming, 108

Yemen, 192–93
Yersinia pestis, 55–56

Zhang Jixian, 12
Zimbabwe, 193
Zinsser, Hans, 81–82